THE TAMING
OF EVOLUTION

THE TAMING OF EVOLUTION

The Persistence of Nonevolutionary Views in the Study of Humans

by Davydd J. Greenwood

Cornell University Press

Ithaca and London

Passages from *On Human Nature* and *Sociobiology*, by E. O. Wilson, and from *Genes, Mind, and Culture*, by Charles Lumsden and E. O. Wilson, are reprinted by permission of Harvard University Press.

Passages from the following books are reprinted by permission of Random House, Inc., and Alfred A. Knopf, Inc.: *The Spanish Temper*, by V. S. Pritchett, copyright © 1954 by V. S. Pritchett; *Cows, Pigs, Wars, and Witches*, by Marvin Harris, copyright © 1974 by Marvin Harris; *Cannibals and Kings*, by Marvin Harris, copyright © 1977 by Marvin Harris; *Cultural Materialism*, by Marvin Harris, copyright © 1979 by Marvin Harris.

Passages from *America Now*, by Marvin Harris, copyright © 1981 by Marvin Harris, are reprinted by permission of Simon & Schuster, Inc.

Passages from *Six Books of the Commonwealth*, by Jean Bodin, translated by M. J. Tooley, are reprinted by permission of Basil Blackwell, Ltd.

Passages and photographs from *Physique and Character*, by Ernst Kretschmer, translated by W. J. H. Sprott, are reprinted by permission of Springer-Verlag.

Passages and photographs from *The Varieties of Human Physique*, by W. H. Sheldon, are reprinted by permission of the Trustees of the Estate of W. H. Sheldon.

First published 1984 by Cornell University Press
Published in the United Kingdom by Cornell University Press Ltd., London.

International Standard Book Number 0–8014–1743–0
Library of Congress Catalog Card Number 84-45147

Printed in the United States of America

Librarians: Library of Congress cataloging information appears on the last page of the book.

The paper in this book is acid-free and meets the guidelines for permanence and durability of the Committee on Production Guidelines for Book Longevity of the Council on Library Resources.

Fried. 24.95 /22.46 |12|3|84

For
Julio Caro Baroja
maestro y amigo

Contents

7

III Complex Continuities 103

Figures

Preface

This book expresses my fascination with certain persistent Western ways of thinking about the relationship between biology and culture. Work that began as an excursion into contemporary nature/nurture controversies soon linked itself to my ethnographic and historical research in Spain and grew into an analysis of the nature/nurture debate as an element in Western culture.

Much is at stake in this controversy. The persistence of the nature/nurture debate, apparently overcoming centuries of scientific attempts to break out of its constraints, tells us much about the ways key cultural notions persist where political and moral ideologies are involved. It also shows how complex the relationship between biological science and culture truly is. We are biocultural animals trying to understand how biology and culture interact and we are consistently derailed by the persistence of certain cultural construals of these relationships.

The project began in 1976, when Cornell University introduced a major called Biology and Society and I was asked to devise a core course on biocultural perspectives for upper-level students with strong backgrounds in the biological sciences. These efforts showed me that there is a small set of recipes for biocultural arguments. The particular branch of biological science involved in the debate matters very little because political and moral issues form the real though often hidden centerpiece. I learned that if one enters this territory on the terms of the combatants, the resulting argument, whether biological determinist or cultural determinist, is both predictable and unrelated to the lessons of evolutionary biology.

Determinist and antideterminist positions also share many more features than would initially seem possible. Concepts of nature and human nature are used in strategically similar ways and the legitimating force of science is appealed to by both sides.

The conjunction of these points suggests a larger issue. In many respects, it appears that the pro- and antideterminist views are not two separate cultural systems in conflict; rather the debate itself is one single cultural system whose central feature is the nature/culture controversy. Scientifically, politically, and morally, these views depend on one another to such an extent that without the loyal opposition, each view dissolves into incoherence. Nature without culture or culture without nature cannot be conceived.

In this work I make extensive use of textual materials interspersed with commentary and analysis. Part I provides a general contrast between pre-evolutionary and evolutionary views of the world through a juxtaposition of humoral theories with the theories found in some of Darwin's major works. Darwin's writings serve as a point of contrast with the pre-evolutionary views and as a set of standards to be applied to contemporary views.

Though one would not think so, this kind of perspective on Darwin is not easy to come by. Darwin suffers from his own celebrity. For most audiences, his ideas are so famous that his works are rarely read. For others, Darwin's worth is measured historically. He provided the first coherent formulation of the theory of evolution that was able to convince a large audience. But "first" also implies "now out of date." Population and molecular biology, ethology, evolutionary ecology, and sociobiology have taken us so far beyond Darwin on so many points that those who view the development of science as a progressive process can see no reason to read his works. Darwin takes his place beside the other great heroes of modern science in a pantheon that is worshiped rather than studied.

Darwin deserves to be reread now, not because he was right about the details of many aspects of the evolutionary process or because he was free of social and moral bias, but because his basic presentation of the structure and requirements of a genuinely evolutionary view of nature has never been bettered. Darwin moved the audiences of his time not simply by his patient accumulation of evidence but by his construction of a view of the world that was capable of overturning much of what was thought before. By comparison, many contemporary statements on evolution, though quite exact about DNA and RNA, are primitive as statements of the evolutionary view.

Part II brings together a diverse set of texts dealing with the question of race, constitutional types, and ethnic and national character. These materials show that pre-evolutionary conceptualizations of the relationship between nature and culture persist in some areas with no modification whatever. This form of persistence is referred to as "simple continuity."

Part III examines two pre-evolutionary views of the natural basis of social order and two contemporary attempts at biocultural synthesis. The first texts center on the idea that social order reflects purity of blood. Others portray an egalitarian social order based on Christian equality before God. These texts are compared with the works of E. O. Wilson and Marvin Harris.

The juxtaposition of Wilson's and Harris' work with the pre-evolutionary texts and those of Darwin provides clear evidence that these contemporary thinkers construct nature and culture in ways that have more in common with the pre-evolutionary texts than with anything in evolutionary theory. Despite my belief in the sincerity of the intentions of both authors, I argue that the claimed evolutionism in their work is an illusion based on the adoption of a few evolutionary terms rather than on the acceptance of the full implications of an evolutionary view of the world. I call this form of persistence "complex continuity."

To make this point, I could have included any of hundreds of works, since these controversies have raged for centuries. I have chosen to examine Wilson's human sociobiology and Harris' cultural materialism because their views have generated such wide interest. Analysis of the works of Ashley Montagu, Robert Ardrey, Konrad Lorenz, Desmond Morris, Anthony Storr, Jacob Bronowski, Arthur Jensen, and many others might have served as well. It was tempting to include the most recent debate in this arena, that between Margaret Mead's supporters and Derek Freeman (Mead 1928, Freeman 1982), but there will always be yet another of these debates to deal with unless the very terms in which they are cast are challenged. I hope to provide such a challenge.

Textual analysis has intrinsic limitations. Texts must be selected from an immense inventory. Assumptions about context and authorial intention are unavoidable. Multiple readings are possible, even necessary. Texts are not determinate phenomena amenable to some transparent technology of interpretation. My analyses of these sets of texts are simply my analyses, to be accepted or rejected on the basis of other readings and juxtapositions with other texts.

Finally, the argument of this work is historical, but only in a limited sense. The comparison of the three major sets of texts is intentionally chronological in order to emphasize the persistence of nonevolutionary ideas in the works of contemporary scholars. A narrative history of these developments would take quite a different form and would be a contribution to the standard genre of the history of science. The obligations implied in the writing of narrative history would distract attention from the points to be made in the comparison of the texts. Should the analysis prove convincing, the narrative history would then have to be written with these points in mind.

Bringing this book to press has been difficult. Because it is sharply critical of both sides of the nature/nurture debate, it appeals to few active participants. I criticize both scholars whose political and moral stances I find objectionable and those whose stances I applaud. Despite the risks, we cannot permit the suspension of intellectual standards simply because the participants believe the issues to be so important that "winning is the only thing." In fact, I believe they are so crucial that we can no longer afford the frivolity of repeating nature/nurture homilies to ourselves.

Pilar Fernández-Cañadas de Greenwood has seen me through this effort with her enthusiasm for intellectual debate, collaboration in research, and detailed editorial suggestions. My son, Alex, understands and supports my passion for these issues.

Walter Lippincott, Jr., director of Cornell University Press, believes that some academic books should be published because they can promote intellectual debate. In an area where peer reviews often yield contradictory advice, he judged that mine is an argument that deserves a public hearing. An author cannot ask for more.

Successive generations of Cornell students in Biology and Society 301: The Biocultural Perspective have had these ideas arrayed before them. Both their enthusiasm and their questions have enabled me to clarify the issues.

Barbara Salazar's expert editing of the manuscript has eased the reader's task. Coraleen Rooney's word-processing skill made my work much simpler.

Many colleagues have been involved with parts of this project. Among those who have offered stimulation and advice are James Boon, William Durham, Sander Gilman, David Holmberg, Edmund Leach, Eric Smith, Bruce Winterhalder, and Aram Yengoyan.

The teachings and friendship of Julio Caro Baroja, to whom this book is dedicated, have been an inspiration to me always.

DAVYDD J. GREENWOOD

Ithaca, New York

THE TAMING
OF EVOLUTION

INTRODUCTION

The Darwinian Revolution?

From the very beginnings of Western thought, ideas about nature and culture and their interrelationships have been central themes in political and philosophical controversy. Moral and political changes have been wrought or reflected in shifting views of nature and culture.

Evolution, as a global theory about the processes that permeate "nature," involves a drastic departure from previous theories and ultimately requires a substantial revision in views of nature and culture. Because the counterposed concepts of nature and culture entail moral and political considerations, evolutionary theory could have had immense social consequences. Social Darwinists and strict creationists were quick to point out this possibility in the nineteenth century, and the continuing struggle between evolutionists and creationists over the teaching of evolution in the public schools shows that the issue is not dead.

Nevertheless, and although evolutionism has indeed revolutionized views of nature, it has dealt successfully with nonhuman nature only. In the study of human beings, the trajectory of evolutionary theory has been obstructed not just by antievolutionists but by many scholars who believe they are applying evolutionary theory to the study of human beings. Evolutionary and antievolutionary views of human nature, though quite incompatible scientifically, can and do coexist.

It will immediately be countered that biological anthropology employs evolutionary principles in the study of humans. No one can doubt that it does. Human origins are generally treated evolutionarily; human genetics and population biology are thriving. Ecological principles are widely applied in the study of human groups, and human ethology is rapidly developing as a formal field of anthropological inquiry. Yet despite anthropology's avowed concern with a holistic biocultural view of humans, biological and cultural inquiry remain surprisingly isolated from each other. The human body, human populations, and human subsistence systems are treated in an evolutionary fashion, while culture is analyzed much as it would have been at any time in Western history before Darwin.[1] This widespread failure to synthesize the study of humans and their cultures cannot be attributed to lack of effort. Numerous attempts have been made, yet as different as their theoretical sources and ideological motivations have been, with very few exceptions they have fallen short in remarkably similar and predictable ways. Thus far, explanations of this general problem have fallen into two broad types: rationality explanations and political-economic explanations. Under the heading of rationality explanations are two subtypes: the "march of science" and unconscious bias.

The "march of science" is seen as an ongoing struggle between the obvious empirical conclusions of science and traditional beliefs. The extent to which evolutionism remains unincorporated in the synthetic study of humans is attributed to a kind of simple survival of prescientific ideas.[2] This view has many weaknesses. It segregates scientific research from other cultural activities in a way that is neither operationally feasible nor anthropologically defensible. It forces us to take a limited view of science as supracultural and to treat the rest of culture outside science as essentially irrational. It also fails to explain what is at stake for the opponents in the conflicts between different views, reducing the issue solely to a battle between superstition and rationality, to warfare between theology and science.

The second kind of rationality view has been nicely synthesized by

Stephen Jay Gould in his fine book *The Mismeasure of Man* (1981). Gould argues that the recurrence of racist uses of IQ tests and other measurement techniques is aided by "unconscious bias." This concept liberates us from the suspicion that all racists are cynical plotters against the truth and it implies the existence of a coherent structure of expectations about the phenomena of the world which guides the thoughts of scientists and nonscientists alike. But unconscious bias is too limited an idea for such a broad explanatory task. To the extent that unconscious biases are shared widely and perpetuated despite use of empirical data and sound analytical procedures, they are not biases at all. They are collective conceptions about the structure and operation of the natural world and its significance for us. They are cultural systems.

The "unconscious bias" explanation individualizes the problem culturally and socially. The persistence of nonscientific views is treated as a matter of individual perceptions, and thus deflects us from an analysis of the evident similarities between the unconscious biases of many scholars and their appeal to a broad public. It also fails to take account of the social interests that are served by particular formulations of the relationship between nature and culture.

Political-economic explanations have been elaborately developed in recent years under the general name "social studies of science." Though still a young field, it offers exciting perspectives on the emergence, acceptance, and rejection of scientific developments.[3] A portion of this scholarship reveals the internal social structures of the biological and medical sciences, traces the social-class interests served by certain of their technologies and ideas, and correlates patterns of theoretical development with political and economic situations. On balance these works suggest—they do not yet demonstrate fully—that the most powerful ideas are generally those that favor the interests of the socially powerful.[4]

Though this is promising work, at its core lies a serious unresolved problem that is recognized by most practitioners. A theoretical perspective that deals with the actual relationship between ideological systems and political-economic structures remains to be developed. When the complexity of individual cases is examined, relatively simple deterministic models prove inadequate. Eventually a theoretical formulation of the relationship between ideology and society capable of taking both ideological and political-economic causes into account will be needed. At present the multiplicity of verbs that are used to refer to this relationship reveals the confusion. Ideological systems are variously said to "reflect," "embody," "correlate with," "describe," "explain," "legitimate," or even "embrace" political-economic structures.

Our ability to deal with all of these problems can be substantially improved if we take a more anthropological view of the way cultural systems operate. The coherence, inclusiveness, and staying power of cultural systems, the interpretive functions they perform, and the moral charge they carry must be brought specifically to bear on the analysis of the use to which evolutionary theory has been put in the study of humans.

This is not a new arena for anthropology, and anthropology is by no means the only actor in it. Since its inception the discipline has taken a central role in polemics about biological determinism, having taken strong public positions against racism, eugenics, and environmental determinism. Anthropologists have argued that what is "natural" to humans is by no means easily determined. Some argue that questions about what is natural to humans are wrongheaded, since all humans become human only through culture.[5] Others claim that with sufficient cross-cultural data and proper analytical care, we may identify human universals that have biological bases.[6] Whichever view one takes, it is clear that the relationship between biological and cultural systems has long been a central concern of anthropology.

Received wisdom suggests that the major difficulty standing in the way of a biocultural evolutionary synthesis is either the incomplete assimilation of evolutionary perspectives in the study of humans or the sheer political/moral manipulation of evidence and theory. Yet most of the current perspectives, both pro- and antideterminist, can be found clearly stated in texts written both long before and long after Darwin's time. There seems to be a characteristically Western way of assimilating information about nature into political/moral views about culture, a way as yet little modified by the development of evolutionary theory.

Many pre- and nonevolutionary views treat species as fixed natural categories that embody the ideal form of each species.[7] This ideal form generally arises in an act of creation that also orders all species into a harmonious system. Creation and subsequent history either are providentially guided or follow some teleological principle.

An evolutionary view, in contrast, treats species as momentary organizations of the immense amount of variation that all organisms produce. These species are formed by natural selection. The dialogue between variation and selection has no inherent direction. The composition of life on earth at any period differs from that in any other period.

It should be clear that pre-evolutionary and evolutionary conceptions of nature are incompatible. Momentary organizations of variation in species cannot be reconciled with ideal and timeless species forms. One must be an evolutionist or not. As obvious as this proposition seems, some scholars think these views to be reconcilable, or at least they throw logic to the winds and reconcile them willy-nilly.

Pre- and nonevolutionary views persist at present in two different ways. In some contexts, pre-evolutionary ideas and language simply continue unmodified. Such thinking is much more prevalent than it may appear. More important, there is a complex form of continuity in which pre-evolutionary terminology has been mostly abandoned but the pre-evolutionary conceptual structure persists essentially intact.

It appears that to most people, pre- and nonevolutionary views are much more attractive than evolutionary views. This attraction requires explanation, especially in light of the great success of evolutionism in the biological and medical sciences. I believe the main reason is that pre- and nonevolutionary views offer a clear relationship between nature and the determination of political/moral conduct. Evolutionism, properly understood, not only explicitly rejects such a relationship but undercuts the vision of nature on which it stands. Evolutionary theory must argue that the difference between "is" and "ought" cannot be bridged by science, a point eloquently made by François Jacob in *The Possible and the Actual* (1982).

Our society characteristically distrusts political and moral systems that do not rest on assertions about nature and human nature. In order to preserve this practice (and its dubious social benefits) we are willing to entertain an amazing amount of contradiction between the evolutionism we claim to subscribe to and the ideological uses we attempt to make of it.

I *Major Western Views of Nature*

To demonstrate that pre-evolutionary views of human nature persist and still dominate thinking about human beings, a necessary first step is to differentiate clearly between pre-evolutionary and evolutionary views of the biological world. In arguments about these issues a clear understanding of the differences between these views is usually assumed to exist; yet the crux of the differences is not clear to a great many of us.

Thus Part I sets out the pre-evolutionary and evolutionary views and tries to sharpen our sense of the minimum requirements that must be met if a view is to be called evolutionary. The crux of the difference between pre-evolutionary and evolutionary views centers on their conception of the categories of things in the biological realm. Pre-evolutionary thinkers asserted the existence of fixed natural categories. Evolutionism construes the biological world as a dynamic system composed of ever-changing species.

No amount of accommodative goodwill can reconcile these two views comfortably. And from these differences arise greater differences in the understanding of the broad historical processes that have resulted in life as we know it and that will continue to operate whether we wish them to or not.

CHAPTER I

Humoral/Environmental Theories and the Chain of Being

Current arenas in the ongoing conflict over the relationship between nature and culture are easy to locate. The terms of debate are familiar because there is a general consensus about the relevant theories and evidence to be discussed. The semantic and lexical impact of evolutionism is strong and provides the needed signposts. But before evolutionism, comparable discussions of the human condition employed different terminologies and styles of argument.

The pre-evolutionary literature dealing with "human nature" and its social, political, ethical, and theological consequences spans the whole Western tradition. Nevertheless, it is difficult to find contemporary analytical discussions that focus directly on these biological issues because of the tendency to dematerialize the material/biological dimensions of important philosophical works and to forget how concerned major thinkers have been with exceedingly mundane problems. The history of biological and medical thought, for example, is often treated separately from the history of Western philosophy.[1]

Naturalistic Views of Society

My choice of materials is guided by some specific assumptions about the necessary contents of any naturalistic view of society, evolutionary or not. Any such view provides answers to the following questions:

1. What exists in the world and how is it organized?
2. How much of the outside world is found inside of human beings?
3. Are humans an entirely special form of creation, or are we constituted of the same matter as all the other things in the cosmos, differing from other life forms only in organization? If it is assumed that humans are composed of the same matter as everything else in the world, a fourth question then arises:
4. How does what is outside in the world get inside of human beings? This question centers on the mixed sources of heredity and environment as they influence human structure and behavior.
5. Is matter from the outside world altered in any way when it is found inside of human beings? This question involves a complex of ideas about the transformation of the primary constituting properties of matter into blood, bone, and tissue as manifested in a human body.
The final question is complexly theological and theoretical:
6. What is left of a human being if all matter from the outside world is removed? Centering on the host of issues about the mind and the soul, this question includes others about the organizational principles that give humans particular characteristics and whether or not these organizational principles are intrinsic aspects of physical matter.

A naturalistic view of society thus provides an explanation of the origin, structure, and behavior of living things and then places humans within that context. It does so against the backdrop of a theory of the structure and operation of the material universe. Darwin's theory of the origin of species and the descent of humans is one such theory.

Before Darwin, a combination of ideas about the material structure of the universe, its relation to the structure and behavior of human beings, and the origin of the entire system dominated Western thought from the fifth century B.C. onward. This tradition includes theories about the physical elements of the universe, their effect on the humoral constitution of human beings, and the origins of the separate categories of living things (i.e., the "origin of species").

A caveat is needed. Over two millennia of Western thought are compressed into a simple mold here. This necessary simplification makes it difficult to see that the attractiveness of these ideas lay precisely in their ability to organize the complex world of experience into intelligible categories for thought and action. A much more elaborate discussion of these issues is available in Ernst Mayr's *Growth of Biological Thought* (1982).

Humoral/Environmental Theories

Humoral/environmental theories form an intriguing and complex explanatory system linking the universe, the earth, humans (as a group and as individuals), and even historical events into a single overall scheme. Simultaneously they provide practical guidelines for conduct. These systems are based on meticulous, consistent observations of the physical, biological, and cultural worlds. They assert that the world is orderly and that it can be studied by systematic means.

The fundamental materialism of this tradition makes a clear appearance in the great medical texts of antiquity. Soon thereafter the separation between material and moral causes of temperament and disease ramifies into a fully developed empirical tradition (Laín Entalgo 1961).

The humoral/environmental theories are always emphatically nonevolutionary. They offer no dynamic, material explanation of the origins of the categories of living things or any continuing process by which new categories come into being. Early on, the species are the work of a variety of gods or processes of interbreeding; later they are produced by the God of Genesis.

The power of these humoral/environmental theories is amply demonstrated by their durability. When in the eighteenth century the Spanish medical thinker Martín Martínez made a thorough critique and attempted renewal of medicine, he expressed his ideas in two volumes of dialogues between a chemist, a follower of Hippocrates, and a follower of Galen. Knowledge of Hippocratic and Galenic texts makes most of Martínez' arguments—and those of his English, French, German, and Italian colleagues—quite intelligible despite the many centuries and scientific discoveries that separate them. Qualities, elements, and humors were as fundamental to eighteenth-century thought as they were in antiquity; these ideas persisted through the Arabic renderings of Galen and other classical writers until the originals were rediscovered during the Renaissance.

The basic structure of ideas consists of sets of dual oppositions. Figure 1 charts their interrelationships. According to their structure of ideas, the material world is divided into primary qualities, elements, and humors. The primary qualities are not themselves manifested directly in matter; rather they are the fundamental characteristics that cause the various compositions of matter in the perceivable world. These qualities are arrayed as two sets of dual oppositions: hot/cold and dry/moist. Everything in the world is a material manifestation of combinations of these primary qualities. The basic elements of perceivable matter—earth, air, fire, and water—result from the differing combinations of the primary qualities. Dry and cold yields earth; moist and cold, water; moist and hot, air; and hot and dry, fire. All things in the material universe are made up of varying combinations of these four elements, which in turn are combinations of the primary qualities.

When these four elements combine to form organic beings, they are converted from elements into humors. These substances that constitute organic beings retain the properties of the elements that give rise to them. In a living body, fire yields yellow bile; air becomes blood; earth becomes black bile; and water yields phlegm. All beings are made up of particular combinations of these four humors. Every species and individual has its particular humoral makeup.

Each of the humors has a direct behavioral counterpart in this complex cultural system. Personal characteristics are explained by reference to the predominance of certain humors and the actions of these humors are explained by their elemental makeup. Direct material causation of physical and behavioral states is thus assumed. The predominance of yellow bile (fire, hot and dry) leads to choleric behav-

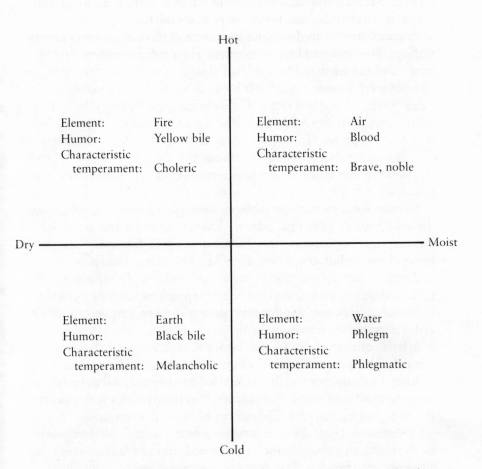

Hot

Element: Fire   Element: Air
Humor: Yellow bile   Humor: Blood
Characteristic   Characteristic
    temperament: Choleric   temperament: Brave, noble

Dry ———————————————————————————————— Moist

Element: Earth   Element: Water
Humor: Black bile   Humor: Phlegm
Characteristic   Characteristic
    temperament: Melancholic   temperament: Phlegmatic

Cold

FIGURE 1 Elements, humors, and characteristic temperaments associated with combinations of the four qualities

ior. The prevalence of blood (air, hot and moist) yields militancy and courage. When phlegm controls (water, moist and cold), passivity prevails; and the dominance of black bile (earth, cold and dry) causes melancholy. Many forces influence the balance of the humors, genealogy and the environment being the primary ones.

In antiquity the study of the humors and their causes and effects (termed *physiologia physis*) embraced the double meaning of "Nature" and the multiple "nature" of things. Nature and the multiple natures were assumed to have a *logos*, a harmonic reasonableness in them. Human *logos* was rational knowledge, among other things. Thus human *logos* attempted to understand and control the *logos* of Nature and of natures. This activity was called *physiologia*—human *logos* seeking to understand and manage the *logos* of the *physis*. And this activity was based on an understanding of the elements and the humors.

To most Westerners, these concepts are vaguely familiar. Even though the language of biles and humors has become obsolete or at least archaic, the complex metaphorical relationships between hot and dry: fire and anger; hot and moist: blood and life; cold and moist: water and peace; and cold and dry: earth and melancholy remain intelligible. Indeed, they are found in ordinary speech to a surprising extent. Powerful literary associations and present-day assumptions about racial characters are based on them.

A basic characteristic of the humoral/environmental theory is its complex hierarchical quality. The universe is constructed of combined primary qualities, physically manifested as elements. All parts of the universe must have these constituents; they must be, in a sense, microcosms of the macrocosm. Differences between distinguishable physical things arise from the quantitative differences in the combinations of the elements of which they are formed. In the organic world the elements are manifested as humors, and the humors, depending on their quality and quantitative mixtures, determine behavior.

From top to bottom and from outside to inside, the universe is made up of varying combinations of four primary qualities. The differences between things reflect only the differing quantitative mixtures of these qualities in the form of elements and humors.

The Environmental and Genealogical Principles

Explanation in this system is basically a matter of accounting for the composition of a particular thing, be it a planet, a rock, a human, a disease, or an emotion. The accounting can be done in two related ways. First, the physical conditions that prevailed at the place and time an entity originated can be used to explain its character—that is, direct environmental causation. For convenience I will call it the "environmental principle" of causation.

Alternatively, the character of an entity can be explained as an inheritance from its progenitors, whose initial traits were fixed through the operation of the environmental principle or by some act of special creation. I will call this the "genealogical principle" of causation. Together these two causal principles play a major role in the social, political, and moral extrapolations based on this theory.

Humoral/Environmental Explanations

These humoral/environmental views are exceedingly adaptable. Major Western developments in cosmology, geography, geology, biology, and medicine reflected the playing out of these basic doctrines for well over two thousand years. Since, in this view, the whole cosmos operated according to material principles that were everywhere the same, certain kinds of scientific research were given major encouragement.

At the same time, these views dictated the terms of major theological questions: What kind of creator would originate such a system and under what circumstances? Could there be anything more to a human being than the combination of four humors? To these questions there are innumerable answers.[2]

Humoral/environmental theories state that the universe is made up of qualities manifested as elements. All four of these elements, transformed into humors, are present in organic benings. Both the environmental principle (immediate physical causation) and the genealogical principle (inherited humoral constitution) are used to explain the way the outside world comes to be part of human beings. Persons are as they are because of where they were born or because of the characters of their parents.

An additional theory is needed to explain how the humors are generated within each human being, and medical history books review a festival of them. Perhaps the most popular view was that the primary humor is the blood: it is the blood that passes on from parent to child, providing the genealogical connection. Accordingly, the blood must somehow contain the other humors or assist in their creation out of elements the person incorporates from the environment. Each generation reproduces itself by means of the blood, which then reproduces the other humors.

Each human being has individual characteristics because of the combination of the humors within. This combination is given by the genealogical principle but is strongly affected by the immediate physical environment, as the subsequent discussion of Hippocrates will show. This dual action of the environmental and genealogical principles neatly sets the foundations of the nature/nurture debate still with us today.

How the parents come to have the particular combination of humors that characterizes them is partly a matter of the blood they received from their parents and partly a matter of their environments. Theorists weight these two factors differently, and arguments over this weighting are a constant feature of these debates. Except in a few extreme cases, so-called hereditarians and environmentalists are separated mainly by a matter of emphasis, a point that deserves more attention than it has received.

Logically implied in this approach to causation is the need for an original creation. The genealogies of entities have to originate somewhere at some time. If one wants to know why humans or for that matter any species has a particular set of characteristics, it is necessary to posit an original moment of creation when those features either first arose directly out of a specific environment or were simply created. Thereafter all history is simply the playing out of combined environmental and genealogical influences as they operate on the original template (constitution, "nature").

The close but ambivalent relationship between the environment and human characteristics is strongly emphasized in this view. Since humans are composed of humors derived from the same elements that make up the surrounding physical/biological universe, humans are easily affected by environmental factors. As the texts of Hippocrates, Galen, and other early medical theorists show, a major part of medical practice had to do with the diagnosis of basic human constitutions and of environmental effects on them, and the elaboration of treatments that altered the internal composition of humors by applications of varying combinations of the elements to the body, either internally or externally.

While the environment leaves it stamp on the individual, the genealogical principle also insists that some of what an individual is comes from the parents, the parents' parents, and so on. Thus some aspects of the individual resist the influence of the local environment. People are as they are because of their ancestors; they "breed true."

This ambivalence between direct environmental causation—a view that is completely consistent with humoral/environmental theory—and genealogical causation, which is much more difficult to assimilate into the theory, is a basic feature of thought in this tradition. Genealogical causation is partly consistent with the theory in that the physical substance of an offspring is directly derived from that of the parent. The massive incorporation of the environment in the form of food and other local influences, however, makes the autonomy of these genealogically transmitted substances hard to maintain theoretically without a well-developed material explanation of heredity. We only now have such an explanation with modern genetics. In most versions of this idea, the genealogical principle is converted quickly into a kind of nonmaterial identity transmitted in the blood.

Given this ambivalence, the central analytical task involved in the application of the humoral/environmental theory to humans is to decide how much of the makeup of a particular person or group of people is caused by the immediate physical environment and how much is caused by the genealogical principle. To the extent that the local environment is a determining factor, a doctor or a ruler can alter the people by altering the environment (insofar as it is possible to do so) or can move the people to a more suitable location.

To the extent that the genealogical principle is a determining factor, a leader can only adjust institutions to suit the characteristics of the people, move them on an environment in which their genealogical traits in combination with the local environment will produce a desired result, or breed them to alter their traits. This is the nature/ nurture argument in its old suit of clothes. Current nature/nurture debates make it clear that little has been added to this view since Hippocrates' time.

The complexity and coherence of the humoral/environmental theory are intriguing. All things—the universe, the planets, animals, humans, individuals—are characterized by varieties of equilibria of their qualities, elements, and humors. The states of sickness and health in individuals vary because the equilibria of their humors vary. Medical treatment acts to restore balance. Individuals differ from one another because of their different personal equilibria. Sexes, races, and species differ for the same reasons that the seasons, periods of life, and periods of history differ. Groups of people in the same places are similar because of their environmentally caused similar humoral makeups. Members of the same families are similar because of their genealogically caused similar humoral constitutions. Thus similarities and differences between people are accommodated and explained by the same small set of principles. Evil, illness, aggression, bad government, and other problems are caused by disequilibria of the elements and humors. Social justice and peace are accomplished by a process of equilibration.

The humoral/environmental theory is thus a very broad theoretical framework for the explanation of the material, organic, and human world. While it cannot account for the initial creation of the categories of entities in the world, it does explain the development and current operation of the world and serves as a guide to conduct. It is thus a theory of great generality and utility.

"Airs, Waters, and Places"

A review of Hippocrates' "Airs, Waters, and Places" (1886) indicates how the theory was actually used as an explanatory model for analyzing human behavior and as a guide to its alteration. Whether or not Hippocrates was a real person is the subject of learned debate.

All that can be said with certainty is that there is a collection of medical texts dating from between 460 and 377 B.C. and that these texts are conventionally grouped under Hippocrates' name.

The famous treatise "Airs, Waters, and Places" is a combined medical handbook and theory of human history based on the analysis of the particular interactions between the elements and their humoral consequences in different environments. It begins with a typology of cities according to the kinds of airs that influence them. Cities exposed to hot winds typically give rise to humid diseases, while those exposed to cold winds produce hard and bilious diseases. The cities in which the west wind predominates combine hot and cold in an unhealthy way that leads to both bilious and febrile disease. Only cities exposed to both hot and cold winds are truly healthy human environments.

A similar typology of waters is elaborated. Marshy areas give rise to bilious diseases. Areas of rock springs cause diseases of hardness. Waters from elevated ground are best, as are waters from the east; those from the west are not bad; those from the north are poor; and waters from the south are extremely unhealthy. Water treatments are recommended for diseases caused by waters. Rainwater, for example, is used to counteract the influence of snow and ice. The treatise also argues that when unlike waters mix, the kinds of airs present will determine which water will prevail. The same system is used to deal with the four seasons. Each season is seen as a combination of temperatures, humidity, and wind—as the working out of the primary qualities in various combinations.

The text then compares Asia with the Mediterranean area in terms of the predominance of various airs and waters and their impact on the human inhabitants. In an overview the author argues that climates that experience great and rapid change give rise to the greatest amount of human diversity, while climates of little change promote human homogeneity.

Up to this point the environmental principle is absolutely paramount. Everything is directly caused by the action of immediate environmental features. But then the text breaks into a different line of argument. The genealogical principle appears and comes into conflict with the environmental principle. In Hippocrates' view, for example, the seasons are modifications of the "natural frame" of the population:

> ... with regard to the country itself, matters are the same there as among all other men; for where the seasons undergo the greatest and most rapid changes, there the country is the wildest and most unequal; and you will find the greatest variety of mountains, forests, plains, and meadows; but where the seasons do not change much there the country is the most even; and, if one will consider it, so is it also with regard to the inhabitants; for the nature of some is like to a country covered with trees and well watered; of some, to a thin soil deficient in water; of others, to fenny and marshy places; and of some again, to a plain of bare and parched land. *For the seasons which modify their natural frame of body are varied, and the greater the varieties of them the greater also will be the differences of their shapes.* [Hippocrates 1886:170–71; emphasis mine]

Having thus raised the possibility of divergence between the local environment and the genealogical principle, the text explores the mechanism by which environmentally caused factors could become part of the "natural frame" in a way that was not definitively challenged until Gregor Mendel's work gained full acceptance.

> I will pass over the smaller differences among the nations, but will now treat of such as are great either from nature, or custom; and first, concerning the Macrocephali. There is no other race of men which have heads in the least resembling theirs. *At first, usage was the principal cause of the length of their head, but now nature cooperates with usage.* They think those most noble who have the longest heads. It is thus with regard to the usage: immediately after the child is born, and while its head is still tender, they fashion it with their hands, and constrain it to assume a lengthened shape by applying bandages and other suitable contrivances whereby the spherical form of the head is destroyed, and it is made to increase in length. *Thus, at first, usage operated, so that this constitution was the result of force: but, in the course of time, it was formed naturally; so that usage had nothing to do with it; for the semen comes from all parts of the body,* sound from the sound parts, and unhealthy from the unhealthy parts. If, then, children with bald heads are born to parents with bald heads; and children with blue eyes to parents who have blue eyes; and if the children of parents having distorted eyes squint also for the most part; and if the same may be

said of other forms of the body, what is to prevent it from happening that a child with a long head should be produced by a parent having a long head? [P. 171; emphases mine]

Humans can be changed by both nature and culture, all within the overall materialism of the humoral/environmental theory. Hippocrates asserts that semen comes from all parts of the body. By this assumption, it is then possible to state that the condition of any part of the body will be reflected in the humoral makeup of the semen that emanates from the part. This makeup, in turn, will be passed on to the next generation in the form of a permanent alteration of the humoral balance, turning cultural practice into a natural feature of a particular group of people.

This is as clear a statement as can be found of the explanation of the path by which external environmental influences become part of the constitutions of individuals. This argument is of crucial importance to later theories regarding the differentiation of races, nations, and constitutional types.

Later in his text Hippocrates shows the way this materialist theory may be reconciled with the existence of a creator:

... there are many eunuchs among the Scythians, who perform female work and speak like women. Such persons are called effeminates. The inhabitants of the country attribute the cause of their impotence to a god, and venerate and worship such persons, every one dreading that like may befall himself; but to me it appears that such affections are just as much divine as all others are, and that no one disease is either more divine or more human than another, but that all are alike divine, for that each has its own nature, and that no one arises without a natural cause. [P. 178]

The natural and the divine explanations are compatible because deities created nature and thereafter it runs according to the principles set in motion upon its creation.

Toward the end of the essay Hippocrates turns to a more detailed analysis of character differences within the Mediterranean world. He argues that in high, varied, well-watered places, people are enterprising and warlike, while in low and ill-ventilated places they are fleshy, dark, and bilious. In high, flat, well-watered places, people are large and gentle, and in high, dry places, people are hard, blond, and haughty. Perhaps the most surprising thing about this last set of observations is how closely they conform to current ethnic stereotypes about regions within modern nation-states and about differences in national character.

These are the basics of the humoral/environmental theory. A complex, tiered system of thought that is immensely adaptable as an explanatory and classificatory framework, it provides the foundation for most of the arguments (and their internal contradictions) regarding relations between nature and culture before and, I believe, after Darwin.

Natural Categories and the Chain of Being

Humoral/environmental theories cannot account unaided for the creation of the categories of entities in the material world. Yet the observable world is filled with highly organized, diverse entities. Each seems to have a "nature"; each represents a species with characteristic structures and habits. Together these species form a reasonably harmonious whole. Why there are many varieties of entities, how they are organized, how they came into existence, the initial source of each, and the degree to which the set of species is complete or in process (progressive or degenerative) are fundamental questions left unanswered by humoral/environmental theories. The answers given to these questions are numerous, but they share certain basic features.

In the span of time separating Hippocrates from Darwin, humoral/environmental theories were associated with various theories of creation and order. Some of the pre-Christian views attribute the categories and order of the material world to the action of a multiplicity of gods. Other conceptions center on single universal creative forces. The most influential view arose from Judeo-Christian thought, in which the God of Genesis became the source of the categories and structure of the material universe.

God creates the natural order. From the initial chaos God orders matter into the elements and then, one by one, fashions all species out of these elements. Further, God creates all beings in a definite system of interrelationship, with humans at the pinnacle. Once this creation is complete, the work is perfect and no further creation is possible. The orderliness, the marvelous structure of even the simplest being, and the complex relations among species all celebrate the divine plan.

The structure of this set of ideas was superbly analyzed decades ago by Arthur Lovejoy in *The Great Chain of Being* ([1936] 1976). This book remains the fundamental work on the subject. Lovejoy's exposition begins with the analysis of a fundamental tension in Plato's thought between other-worldliness and this-worldliness. By "other-worldliness" Lovejoy means not a belief in an afterlife but rather

> the belief that both the genuinely 'real' and the truly good are radically antithetic in their essential characteristics to anything to be found in man's natural life, in the ordinary course of human experience, however normal, however intelligent, and however fortunate . . . the human will, as conceived by the other-worldly philosophers, not only seeks but is capable of finding some final, fixed, immutable, intrinsic, perfectly satisfying good. . . . Not, however, in this world . . . but only in a 'higher' realm of being differing in its essential nature, and not merely in degree and detail, from the lower. [Lovejoy 1976:25–26]

This higher realm is the world in which the multiplicity of eternal Ideas resides, ideas of which the things of the natural world are but imperfectly realized copies. Lovejoy goes on to show the basic contradiction between devaluing the natural world and using it as an empirical source for the formulation of the eternal Ideas.

All categories of things in the natural world are imperfect manifestations of perfect and eternal Ideas, but these Ideas are not randomly organized. They are ordered in turn by an "Idea of the Idea," generally rendered by Plato and eventually by Christian theologians as the "Idea of the Good." Lovejoy handles it deftly:

> . . . it is . . . the most indubitable of all realities . . . it is an Idea or essence . . . in distinction from the particular and changing existences which in varying degrees participate in its nature; and . . . it therefore has the properties common to all the Ideas, of which the most fundamental are eternity and immutability . . . it is the polar opposite to 'this' world . . . its true nature is therefore ineffable in the forms of

ordinary speech . . . the Form of the Good is the universal object of desire. [Pp. 40–41; emphasis removed]

If this idea is accepted, then, as Lovejoy points out, two problems are left unsolved. First, it is not clear why the imperfect natural world should exist at all. And second, if the natural world must exist, what determines how many and what types of things must make it up? The answers to these questions set the relevant point of comparison between pre-evolutionary and evolutionary views of the organic world.

To explain why the natural world must exist, theologians argued that the perfectly good cannot envy anything not part of itself.

Its reality could be no impediment to the reality . . . of beings other than it alike in existence and in kind and in excellence; on the contrary, unless it were somehow productive of them, it would lack a positive element of perfection, would not be so complete as its very definition implies that it is. . . . The concept of Self-Sufficing Perfection, by a bold logical inversion, was . . . converted into the concept of a Self-Transcending Fecundity. [P. 49]

Once this idea is accepted, the question regarding the number of things the world must contain is easily answered. Since the perfectly good could not envy the existence of anything, the world must contain absolutely all possible kinds of things. The world must be completely filled with all the possible natural categories of things. This concept Lovejoy christened the "principle of plenitude."

And once the principle of plenitude is accepted, the question of the organization of things in this world is also easily resolved. Since the perfectly good is the universal object of desire, and the world of nature is an imperfect copy of it, then all things in the world are arrayed in a hierarchy from the lowest and most distant from the perfectly good to that which is closest to it. In addition, the principle of plenitude requires that there be no gaps in the hierarchy from the lowest to the highest, since the world would thus have less in it than it could have. This hierarchy of categories of things, animate and inanimate, is what came to be known as the "chain of being."

The humoral/environmental theories provide explanations about the way entities in the material world behave through analysis of their material constitutions. On this level the humoral/environmental explanation is consistently materialist, since the same principles are applied to molten lava and epileptic seizures. But the humoral/environmental theory cannot explain the existence of the categories of things in the world or the origin of their interrelationships. Each kind of stone, plant, animal, air, water, human, disease has its source in principles external to the humoral/environmental view. The source is the eternal Ideas of which the things of the world are only manifestations. Only when humoral/environmental theories are armed with this conception can they become a global explanatory framework.

Let us return for a moment to the contrast between the genealogical and environmental principles. At first it appears that humoral/environmental theories rest primarily on the environmental principle, but it quickly becomes apparent that the genealogical principle is equally important, because only through genealogy can the categories of entities in the world and their interrelationships be explained.

As a result, humoral/environmental theories are always riven by a conflict between the environmental and the genealogical principle. The ambivalence can be seen in Hippocrates' discussion of the long-headed Macrocephali and in his concept of the "natural frame" that is resistant to environmental influences.

The essences of things, the categories of the natural world, are given by the eternal Ideas. They pass from generation to generation by the genealogical principle. Their transmutations occur only as the result of the action of humoral/environmental principles, and these transmutational principles themselves are ultimately derived from eternal Ideas.

The conflict between the necessary genealogical principle and the primarily environmental bias of the humoral/environmental explanations can never be resolved. The nature/nurture debate is an intrinsic part of this view. It is not, however, intrinsic to modern evolutionary biology, and the persistence of the nature/nurture debate up to the present suggests just how powerful the hold of the ancient humoral/environmental world view still is.

CHAPTER 2

Evolving Natural Categories: Darwin's Unique Legacy

The evolutionary world is a world in motion, in stark contrast with the static world of Hippocrates and the chain of being. At the very center of evolutionary thought is a view of the classification of living things so radically different from earlier views that it turns them inside out.

Though empirical investigation made a major contribution to the development of evolutionism, the Darwinian revolution cannot be understood as the simple result of such activities. Darwin's own view of the organic world seems to rest on a systematic inversion of the structure of earlier views, an inversion that then served as the source of his hypotheses for empirical investigation. Any approach that claims to be evolutionary must share certain elements of Darwin's vision of the world in motion. Many putatively Darwinian views do not.

The scholarship on Darwin and Darwinism is enormous and rich.[1] The intense debates about the dating and sources of ideas are pursued by experts with lifelong commitments to the study of Darwin's texts. Radical reperiodizations of Darwin's thinking are currently taking place, particularly in light of the information contained in his notebooks

(Gruber, ed., 1974; Herbert, ed., 1980). The young Darwin had developed the whole theory of classification and evolution much more fully than anyone reading *The Origin of Species* would imagine. Darwin's representation of himself as a mere accumulator of data that inexorably led him to the theory of evolution is now being questioned.

Despite the appreciation of Darwin's own work and general agreement on the revolutionary impact of evolutionism on biological, social, and moral thought, one crucial element is widely mentioned but still tends to be undervalued. Darwin proposed a radical revision in our understanding of taxonomic classification in biology and in the theoretical meaning of the various taxa within the system. Substituting genealogical for formal principles of classification, Darwin reorganized the system of classification to reflect descent with modification and to emphasize the productive role of variation in evolutionary processes.[2]

Culture and Classification

The study of systems of classification is a major component of modern sociocultural anthropology. Beginning early with the study of the alternative means by which people in other cultures classify the world into "kin" and "nonkin" (thereby structuring their social relations in crucial ways), more recently passing on into the study of totemic classification systems (Douglas 1966) and the development of the entire field of structuralism and semiotics (Lévi-Strauss 1962), and finally into the study of ethnoclassification (Berlin and Kay 1969), anthropologists have devoted immense energy to the analysis of systems of classification. In particular, structuralist and semiological approaches to the analysis of cultural systems as systems of classification have yielded fascinating results in the hands of Claude Lévi-Strauss, Mary Douglas, Edmund Leach, and many others.[3]

One of the most important results of these anthropological analyses is an increasing awareness of the close linkage between systems of classification and systems of morality. Any alteration in the major classifications of a cultural system nearly always implies alterations in moral systems as well. Put another way, there are no morally insignificant classifications.[4]

Seen in this light, the explosive reaction to Darwin, including the bitter questioning of his moral character, is not surprising. In proposing a revolution in the system of biological classification, evolutionism demolished part of the foundation of existing moral systems. This conflict accounts for the persistence of the essentialist/populationist debate in biology that Ernst Mayr (1982) so adroitly documents.

No evolutionist has stated the case better than Darwin did. It was his particular genius to have linked the question of evolution tightly to the issue of classification. But Darwin is no culture hero. He was wrong about many things, inconsistent about others; significant social biases invade his work. Many commentators, including Karl Marx, have claimed that Darwin's view of the evolutionary process was strongly conditioned by his social experience and position; that he was, in a word, an apologist for the capitalist order. Clear lines of evidence can be traced to show that Darwin held important social, racial, and sexual biases, though one becomes impressed by the moderation of his positions on these issues as one reads the works of many of his contemporaries.

Darwin was very much a man of his era, sharing many of its ideological biases. Yet no one has ever made the case for evolution better or with a clearer recognition of the immense scope of the intellectual revolution it proposes. In Darwin's key texts we are able to see clearly the revolution in classification that evolutionism implies. Perhaps better than any of his contemporaries and despite his ideological biases, Darwin believed that political and moral conclusions could not be derived directly from the study of evolution, a view exceptional even now.

Precisely because Darwin's works are so important, virtually every aspect of them has been subjected to close scrutiny. Much of this territory is hotly contested. Darwin's theories themselves, his social biases, his religious beliefs (or lack of them), and many other aspects of his writing and character are debated by experts. Little can be said about Darwin that will not be repudiated by someone. Such is his fate.

Darwin's Major Works

The evidence Darwin used was available to all, and many of the conceptual structures were also in wide circulation. The secret of Darwin's success goes beyond his theoretical and empirical innovations. After all, both Darwin and his colleague Alfred Russel Wallace had come up with the theory of evolution. But it was through the writing of *The Origin of Species* that Darwin began the reversal of the predominant Western system of classifying nature and the construction of the theory of evolution out of that reversal. My emphasis on the reversal accounts for the highly selective reading of Darwin presented here.

I make no claims regarding Darwin's personal intentions. I do not know the thought process by which Darwin constructed his works. Rather I emphasize the internal logic of his system of thought as it supports the theory of evolution. The test of this kind of approach is simply how well it accounts for the structure, emphases, and data arrayed in the works.

The Origin of Species

Virtually everyone knows the story of the writing of *The Origin of Species*: the voyage on the *Beagle*, the coincidence of Darwin's and Wallace's views on evolution, the great stir Darwin's book created, and Darwin's subsequent problems and doubts. These issues are left aside here to concentrate on some of Darwin's statements about classification of plants and animals. Here his views contrast starkly with those of his predecessors and with those of many of his supposed followers as well.

The very title bears close scrutiny: *The Origin of Species by Means of Natural Selection; Or the Preservation of Favoured Races in the Struggle for Life*. The book is about the *origin* (in the sense of productive principle and genealogy), not the character, behavior, morphology, or embryology of species. It is about the origin of *species*, that is, how the categories of living things in the world of observation have come into existence—a process until then explained by the idea of special creation, which organized species into a chain of being, the species then being acted upon by humoral/environmental forces. The term "natural selection" is used because the causal agent in speciation is not God or any other supernatural principle that at some time in the remote past created eternal categories; it is the continuous operation of blind material laws.

The restatement of the subject—*Or the Preservation of Favoured Races in the Struggle for Life*—brings up an ambiguity that runs through Darwin's work and that of most other nineteenth-century thinkers: the terms "species" and "race" are often used interchangeably. In Darwin's case, the use of the term "race" is important because in ordinary speech "race" conveys a strong sense that genealogical principles are the primary ones that give rise to and define species boundaries. Species are separate categories of things in the world. To call them races is to emphasize the genealogical lines stretching from past to present and set the scene for the idea of descent with modification.

The "struggle for life" notion points to reproductive competion under specific environmental conditions as the mechanism by which natural selection operates. It is not a providential hand that chooses but blind physical laws, and the struggle never ends. Thus Darwin titled his book well, for the title summarizes his theory.

The book itself, despite Darwin's continual reference to it as an essay written in haste, has a coherent internal structure that supports the argument well. It is divided into three major setions, each dealing with a specific set of issues.

The first section is devoted to an exposition of the mechanisms that produce variation, competition, and ultimately speciation by means of natural selection. Darwin begins this section strategically by discussing variation under domestication. Since humans have induced speciation quite often by manipulating the environments and reproduction of plants and animals, how can we believe that species boundaries are absolute and fixed? Darwin also points out that species under domestication are highly variable. It is this variability that permits us to breed plants and animals as we do.

Variation is the centerpiece of Darwin's world view, and thus he begins with it. Where others saw uniformity and clear-cut species boundaries, Darwin saw ranges of variation. Much of his observational life was spent in cataloguing variation, a phenomenon that not only was theoretically important for him but clearly fascinated him in its own right. This focus contrasts sharply with the pre- and non-evolutionary views of the world, in which fixed natural categories of things succeed themselves from generation to generation, with no change other than some preordained progress or fateful degeneration.

Having made his argument about variation and selection in the world of everyday human experience, Darwin then moves outward to the undomesticated world, showing that a similarly great amount of variation exists in all living things. Examples abound, drawn from everywhere. Variation is the first step in the argument for evolution.

Darwin then introduces the "struggle for life." Outside the domesticated world there are no plant and animal breeders to act on variation. Darwin argues that different variants are differentially capable of reproducing successfully in the diverse environments in which they are found. Those better able to reproduce proliferate and the others fade away. This view is complemented by the naming and development of the concept of natural selection.

In Chapter 1, humans were the selectors. Now Darwin opens up the possibility that blind natural forces act as selectors on the immense number of variants among plants and animals. This process of natural selection gradually changes the structure of species of plants and animals until new and different species come into existence, all genealogically linked to their predecessors. He closes the section with a chapter on "laws of variation," in which he claims that a theory based on variation and natural selection will yield a coherent interpretation of the history of life on earth.

The second section strategically poses the major objections that could be advanced to compromise his general argument. Here Darwin's capabilities as a scientific thinker become particularly manifest as he first conceptualizes the opposing views and then systematically undercuts them. He casts down the gauntlet in scientific fashion: this is the theory; here are all the objections to it; none of the objections is crippling, so until an unanswerable objection can be formulated, the theory stands.

In the third section Darwin arrays geological, geographical, and morphological evidence in support of the theory of evolution. With each body of evidence goes the suggestion that much more evidence could be adduced. The recapitulation of the whole argument at the end of the book shows, among other things, just how well planned this "essay" was. Throughout the book the movement is measured and self-conscious: argument, counterargument, proof, recapitulation.

Often using little and tenuous evidence, Darwin succeeded in making a forceful argument that species arise continuously by material processes, thereby demolishing the centerpiece of the pre-evolutionary views of nature. Darwin did not create an unrecognizable world of nature. Rather he took the world of nature as people observed it and explained its structure in a way that was odious to a significant part of his audience.

It is helpful to flesh out the argument about evolutionary classification put forward here with some of Darwin's specific statements. *The Origin of Species* begins with a criticism of the view of species as immutable and separate:

> Until recently the great majority of naturalists believed that species were immutable productions, and had been separately created. . . . Some few naturalists . . . have believed that species undergo modification and that the existing forms of life are the descendants by true generation of pre-existing forms . . . [Darwin (1859) 1958:17]

An emphasis on genealogical relationships (what Darwin called "true generation") is the key to his argument against separate creation. Occasionally this emphasis on genealogical relationships leads him to appeal to the old language of "blood," as in the following case: ". . . looking to the domestic dogs of the whole world, I have . . . come to the conclusion that several wild species of Canidae have been tamed, and that their blood, in some cases mingled together, flows in the veins of our domestic breeds" (p. 39). He contradicts the notion of separately created species by conjuring up genealogical relations between wild and domesticated dogs. If such genealogical relations exist, and the wild comes before and gives rise to the tame, then species are generated out of each other.

Later Darwin links variation with speciation, in this case by speaking of domesticated species: "The key is man's power of accumulative selection: nature gives successive variations; man adds them up in certain directions useful to him. In this sense he may be said to have made for himself useful breeds" (p. 48).

This is a frontal attack on special creation. Nature itself is the producer of variation. In the pre-evolutionary view, nature merely reflected the initial order made by the creator. Humans produce species by acting on this variation; humans "create" species through selection. Such actions would be impossible in a world of hermetically sealed separate species. For Darwin, variation is not deviation from the ideal form of the species; it is the source of all species.

Darwin sums up this argument by stating:

> Changed conditions of life are of the highest importance in causing variability, both by acting directly on the organisation, and indirectly by affecting the reproductive system. . . . Over all these causes of Change, the accumulative action of Selection, whether applied methodically and quickly, or unconsciously and slowly but more efficiently seems to have been the predominant Power. [P. 57]

Thus by the end of his initial chapter Darwin has put most of the major elements of his argument into play. Beginning with the world of everyday observations of domesticated plants and animals and their variations, he moves outward toward larger principles. If humans, by controlling reproduction, can create species by selecting for certain variants, then species are not immutable. It is then conceivable that other agencies (such as the environment) could also select among the variants within an existing species and ultimtely cause the creation of new species. If this argument is accepted, separate creation and the chain of being disappear.

The central obstacle left for Darwin to overcome is the proof that variation and selection occur outside of the domesticated species directly subject to human agency. He turns to this task with vigor.

The argument begins with a strong emphasis on the arbitrariness of species boundaries. Darwin states: ". . . I was much struck how entirely vague and arbitrary is the distinction between species and varieties" (p. 63). Then he elaborates the argument:

Certainly no clear line of demarcation has as yet been drawn between species and sub-species ... or, again, between sub-species and well-marked varieties, or between lesser varieties and individual differences. *These differences blend into each other by an insensible series; and a series impresses the mind with the idea of an actual passage.* [P. 66; emphasis mine]

Darwin's view of the empirical world centers on variability, complexity, and classificatory ambiguity. He radically reverses the view of nature as a linked chain of sharply separable species with no motion other than self-replication. In place of fixed species Darwin inserts "an insensible series." In place of clear categories he interjects the arbitrariness of species boundaries. And in place of a series of similar but variant organisms he poses a historical, evolutionary relationship.

To counter the existing view of the natural world, Darwin attacked the dominant classificatory scheme. Rather than seeing species as eternal categories that empirically vary around the perfect expression of the species' inherent character, he makes variation the real and eternal feature of nature, converting species into momentary historical embodiments of these variations. In this massive reversal he subverts special creation and the chain of being at the same time. He creates a view a nature that is the opposite of the dominant one.

This argument logically leads him to the next questions: What agency causes the appearance of various species out of the multiplicity of variation observed in nature? Why is nature simply not an incoherent teeming of life? Many thinkers have attempted to reinsert the hand of God into the theory of evolution at this point by arguing that the whole process of variation and selection is providentially guided. Darwin did not take this path. Instead he insisted that the origin of species must be understood solely as a historical process with a beginning, operating according to coherent principles, but without an intrinsic direction or goal.

Defining the struggle for life, he says:

Owing to this struggle, variations, however slight and from whatever cause proceeding, if they be in any degree profitable to the individuals of a species, in their infinitely complex relations to other organic beings and to their physical conditions of life, will tend to the preservation of such individuals, and will generally be inherited by the offspring. The offspring, also, will thus have a better chance of surviving, for, of the many individuals of any species which are periodically born, but a small number can survive. I have called this principle . . . by the term Natural Selection, in order to mark its relation to man's power of selection. But the expression often used by Mr. Herbert Spencer of the Survival of the Fittest is more accurate. . . . [P. 74]

Variation, complexity, multiplicity, and competitive advantage are the centerpieces of Darwin's view.

Why could God not create such a world? Darwin does not, in fact, dispute the existence of God, but he deeply distrusts human understanding of nature. Darwin's view of the physical limitations of the human mind and its capacity for experience in relation to the complexity and vastness of the physical universe makes him deeply suspicious of simplistic conceptions of the creator.

> . . . so profound is our ignorance, and so high our presumption, that we marvel when we hear of the extinction of an organic being; and as we do not see the cause, we invoke cataclysms to desolate the world, or invent laws on the duration of the forms of life! . . . He who believes that each equine species was independently created, will, I presume, assert that each species has been created with a tendency to vary, both under nature and under domestication, in this particular manner, so as often to become striped like the other species of the genus; and that each has been created with a strong tendency, when crossed with species inhabiting distant quarters of the world, to produce hybrids resembling in their stripes, not their own parents, but other species of the genus. To admit this view is, as it seems to me, to reject a real for an unreal, or at least for an unknown, cause. It makes the works of God a mere mockery and deception; I would almost as soon believe [with] the old and ignorant cosmogonists, that fossil shells had never lived, but had been created in stone so as to mock the shells living on the seashore. [Pp. 82, 155–56]

By emphasizing the diversity and complexity of the natural world, Darwin stresses a point he makes repeatedly in the *Origin* and later, in more detail, in *The Descent of Man*.

Darwin is very careful to argue that natural selection itself does not induce variation. Were he to admit that it did, variation would be easily reduced to a mere incident in natural selection, with evolution following directly from environmental requirements. As it is, his emphasis is just the opposite. There is hugely more variation available in the world than is operated on by selection. Variation is produced by principles that Darwin did not in fact understand, but he was certain that the cause was not the process of selection itself. Living nature's prime characteristic is production of variation; the production of temporary categories (species) is a subsequent step. The world moves dialectically from variation to selection to variation. Selection creates coherent classes of organisms.

Darwin argues that the crossing of varieties is a law of nature itself. If the chain-of-being argument were correct, then logically the best examples of a species would fertilize themselves. But Darwin's observations lead him

> to believe that it is a general law of nature that no organic being fertilises itself for a perpetuity of generations; but that a cross with another individual is occasionally . . . indispensable. On the belief that this is a law of nature, we can . . . understand several large classes of facts . . . which on any other view are inexplicable. [P. 101]

Reproduction is mixing of variants.

Darwin reverses separate creation by considering the variations to be real and continuous and the species to be only momentary manifestations. He then argues that species breed in such a way as to produce rather than to reduce variation. Thus variation, not the fixity of species, is the "law of nature."

Darwin has chosen his words well, breaking the old meaning of "law of nature" from its mold as a description of a fixed order. He moves on a chain of rhetorical links from variation to domestication, breeding, and natural selection. At each juncture he challenges others to explain these facts better. Nature is a complex set of processes, a history with no goal. This is what Darwin's chart of species change emphasizes. (See Figure 2.)

When Darwin discusses the geological record, he strongly states his historical view of nature as in continual motion against the view of nature as consisting of a set of fixed, uniform categories. The constancy in nature is a constancy of motion:

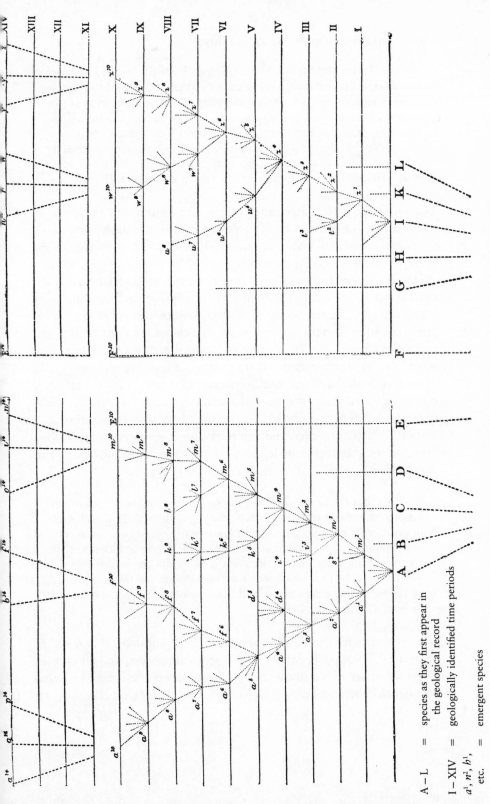

FIGURE 2   Darwin's diagram of the evolutionary process (Darwin [1859] 1958)

A – L   =   species as they first appear in
               the geological record

I – XIV   =   geologically identified time periods

a^1, n^2, b^1,   =   emergent species
etc.

We can clearly understand why a species when once lost should never reappear, even if the very same conditions of life, organic and inorganic, should recur. For though, the offspring of one species might be adapted . . . to fill the place of another species in the economy of nature, and thus supplant it; yet the two forms—the old and the new— would not be identically the same; for both would almost certainly inherit different characters from their distinct progenitors; and organisms already differing would vary in a different manner. [P. 315]

The uniformity of causes imparts order to nature, but these causes work only through time. Each successive moment in the history of life is different from all other moments. This is a direct attack on the standard view of special creation.

Darwin shows his awareness that the issue of classification is the central one in evolutionary theory when he points to "the difficulties which are encountered on the view that classification either gives some unknown plan of creation, or is simply a scheme for enunciating general propositions and of placing together the forms most like each other" (p. 387). He points out that classifications are a complex mix of morphological and functional criteria; that embryological affinities not obvious in adult members of species are important in classifications; and that geographical distributions have also been important. Darwin insists, however, that there is only one correct overarching principle of classification. It must be followed, with all other principles subordinated to its requirements:

I believe that the *arrangement* of the groups within each class, in due subordination and relation to each other, must be strictly genealogical in order to be natural; but that the *amount* of difference in the several branches or groups, though allied in the same degree in blood to their common progenitor, may differ greatly, being due to the different degrees of modification which they have undergone; and this is expressed by the forms being ranked under different genera, families, sections, or orders. [P. 391; emphases his]

Note that he says the arrangement must be "genealogical in order to be natural." Nature is defined as a historical process, and all classifications that are to aid in clarifying this natural process must be based on historical relationships. Darwin does not deny that classification can be done on numerous grounds, but he does deny that any but a genealogical classification can be used as the basis of an evolutionary analysis.

There is no question that the reaction to *The Origin of Species* was appropriate. The shock, excitement, and outrage were fully earned by such a frontal attack on the basis of existing views of nature. The *Origin* is a polemic. Despite its often ponderous apparatus of facts and its flat language, its central structure is provided by a reversal of the Western view of nature, supported by an appeal to scientific appraisal of the "facts," to the value we attach to the scientific method and rational analysis.

The principal issue he did not attack directly was that of human evolution. The only reference to it is the famous cryptic line "Much light will be thrown on the origin of man and his history" (p. 449). But no one could doubt what Darwin had in mind. If special creation and the chain of being had given humans a privileged place in nature, then a repudiation of those views must also alter the human position. It did, though his full statement of the point was not published until 1871, under the title *The Descent of Man and Selection in Relation to Sex*.

The Descent of Man and Selection in Relation to Sex

The *Descent* is a remarkable work both for the currency of its arguments and for the relative obscurity into which it has fallen. While The *Origin* is available in a variety of popular editions, until 1981 the *Descent* was for a long time available only in expensive facsimiles. Thus its readership has been quite limited, though the general public has made best-sellers of other books on the same general topic (Lorenz [1963] 1971, Morris 1969, Ardrey 1966).

Evolutionary analysis applied to humans and the theory of sexual selection appear together in the *Descent* because Darwin believed that natural selection was much relaxed in the human case. In its place, sexual selection, heightened by human cultural capacities for its elaboration, is seen as a prime force. Darwin uses sexual selection to explain, among other things, the phenotypic diversity of human "races."

At each point in the long argument, humans are shown to be influenced by the same forces that are at work in the rest of the natural world. Thus the *Descent* closes out the hope that a unique realm of natural laws might be preserved that would maintain the special dignity of human beings.

Darwin emphasizes human variability just as strongly as he emphasized the variability of nonhuman life in the *Origin*. "The variability or diversity of the mental faculties in men of the same race, not to mention the greater differences between men of distinct races, is so notorious that not a word need here be said. So it is with the lower animals" ([1871] 1974:26).

After making some remarkably contemporary-sounding comments to the effect that humans are like domesticated animals in our variability, in the way selective forces affect us, and in that we are a highly diverse, wide-ranging species, he argues that human superiority is a direct result of our success in the struggle for existence:

> Man in the rudest state in which he now exists is the most dominant animal that has ever appeared on this earth. He has spread more widely than any other highly organized form; and all others have yielded before him. He manifestly owes this immense superiority to his intellectual faculties, to his social habits, which lead him to aid and defend his fellows, and to his corporeal structure. The supreme importance of these characters has been proved by the final arbitrament of the battle for life. Through his powers of intellect, articulate language has been evolved; and on this his wonderful advancement has mainly depended. [Pp. 46–47]

After these generalities, Darwin focuses the argument on each of these traits in turn: intellectual faculties, social habits, body structure. Regarding mental faculties he says:

> My object . . . is to show that there is no fundamental difference between man and the higher mammals in their mental faculties. . . . As no classification of the mental powers has been universally accepted, I shall arrange my remarks in the order most convenient for my purpose; and will select those facts which have struck me most with the hope that they may produce some effect on the reader. [P. 64]

Regarding social living, Darwin argues that social habits and social control depend on the animal's ability to sense the approval and disapproval of its conspecifics. He calls this capacity "sympathy" and claims that it also evolved by natural selection. All social animals have it, but humans have it in a higher degree. Darwin thus erases the line between human and nonhuman.

In the course of his argument Darwin explores the notion that different environments act as spurs to different kinds of cultural systems, especially the idea that challenging environments stimulate great activity and industriousness. He advances a number of eugenic arguments as ways of seeing natural selection at work on the populations of "civilized" societies, though his position is moderate by comparison with Francis Galton's hard line ([1869] 1962). According to Darwin, the instinct of sympathy—the instinct to act in accordance with our understanding of the effect our actions will have on other people—is the basis of our social existence. It prohibits us from taking positive eugenic measures against the weak.

After delivering the judgment on the animal affinities and genealogy of humans for which Darwin is so famous, he attempts to replace absolute human superiority with a sense of the majesty of life itself:

> Thus we have given to man a pedigree of prodigious length, but not, it may be said, of noble quality. The world, it has often been remarked, appears as if it has long been preparing for the advent of man: and this, in one sense is strictly true, for he owes his birth to a long line of progenitors. If any single link in this chain had never existed, man would not have been exactly what he now is. Unless we willfully close our eyes we may, with our present knowledge, approximately recognize our parentage; nor need we feel ashamed of it. The most humble organism is something much higher than the inorganic dust under our feet; and no one with an unbiased mind can study any living creature, however humble, without being struck with enthusiasm at its marvelous structure and properties. [P. 161]

When Darwin takes up the vexed question of race, he remains true to his principles. He begins by returning to the general argument that the question of race must be very ambiguous if even species boundaries are fuzzy. He then states that the question whether the races are species or not cannot be solved in the absence of a general definition of species that is accepted by all biologists. Lacking such a definition, he argues that although the existing races of man differ in many respects ... yet if their whole structure be taken into consideration they are found to resemble each other closely in a multitude of points. Many of these are so unimportant or of so singular a nature, that it is extremely improbable that they should have been independently acquired by aboriginally distinct species or races. The same remark holds good with equal or greater force with respect to the numerous points of mental similarity between the most distinct races of man. ... The great variability of all the external differences between the races of man, likewise indicates that they cannot be of much importance; for if important, they would long ago have been either fixed and preserved, or eliminated. In this respect man resembles those forms, called by naturalists protean or polymorphic, which have remained extremely variable, owing, as it seems, to such variations being of an indifferent nature, and to their having thus escaped the action of natural selection. [Pp. 174, 193]

To account for the great differences among the races, Darwin turns to the principle of sexual selection. He views the development of highly distinctive morphological and behavioral characteristics primarily as means of attracting mates.

The work concludes with the following summary:

The main conclusion here arrived at, and now held by many naturalists who are well competent to form a sound judgement, is that man is descended from some less highly organized form. The grounds upon which this conclusion rests will never be shaken, for the close similarity between man and the lower animals in embryonic development, as well as in innumerable points of structure and constitution, both of high and of the most trifling importance ... are facts which cannot be disputed. They have long been known, but until recently they told us nothing with respect to the origin of man. Now when viewed by the light of our knowledge of the whole organic world, their meaning is unmistakable. The great principle of evolution stands up clear and firm, when these groups of facts are considered in connection with others, such as the mutual affinities of the members of the same group, their geographical distribution in past and present times, and their geological succession. It is incredible that all these facts should speak falsely. *He who is not content to look like a savage, at the phenomena of nature as disconnected, cannot any longer believe that man is the work of a separate act of creation.* [Pp. 601–2; emphasis mine]

The brutality of his last line is stunning. The choice we are left with is to be primitives and believe in special creation or to be civilized and try to make do with evolution and find new sources of species pride.

At the very end the book adopts an uneasy balance between a view of our social obligations to educate people as far as possible (because increasing intellectual awareness necessarily improves moral judgment) and Darwin's laissez-faire view of society as an arena of free social competition. The book contains many tensions of this sort. There are lapses into racist doctrines; there is a good deal of overt and covert sexism; eugenics is toyed with. Yet by comparison with such contemporaries as Spencer and Galton, Darwin was very cautious on these issues—more ready than they to accept responsibilities for the protection of the weak than to use evolutionary doctrines to justify the suppression of the poor and the defenseless among us, and less willing to derive his ethics from the study of biology.

Evolving Natural Categories

My contentions are simple, perhaps even uncontroversial. I have argued that Darwin was a true scientific revolutionary. If one takes the view of nature (and of humans) that supports the theory of special creation and the concept of the chain of being and reverses all their central postulates, one comes up with a view very much like Darwin's. Whether or not Darwin consciously engaged in such an inversion in immaterial. What matters is that conceptualizing Darwin's work this way permits us to understand its revolutionary impact and to set the minimum requirements that any view of humans that claims to be evolutionary must meet.

The theories of special creation and the chain of being demand acts of creation that give rise to all categories of living things. These categories have definite natural boundaries. What variation there is among them is due solely to the influence of environments and, in the case of humans, to sin. Nature has no history, if history is conceived as a continual, open-ended causal process. At most an idea of progress or degeneration can be inserted in an attempt to make history dynamic without doing violence to the idea of special creation.

The special-creation view requires a fixed, absolute hierarchy of separately created species. The order of hierarchy is set by the sequence of creation laid out in Genesis. At the end of the sixth day of creation there is a day of rest. A radical break occurs when Adam and Eve are expelled from the Garden of Eden; now secular historical time begins. But this secular time has no capacity to bring about any alteration in the separately created categories. Each and every creature in every successive time period is simply the reincarnation of its species as that species was originally created. (See Figure 3.) To the extent that there is a genealogical relationship between time periods, it simply stretches back to the first created individual of the species. Any variability observed in any of the time periods is seen as deviation from the created ideal, deviation caused by environmental effects or by degeneration.

In this view, the core element is the idea that plants and animals breed true. As a first principle, humoral/environmental theories assert that there are direct physical principles of causation that operate in harmony throughout the physical universe. The second principle is that a first creation had to occur in a specific environmental context and this creation gave rise to the species constitutions that thereafter have passed through time almost intact. Reproduction in this view means copying. The environmental and genealogical principles interact by special creation, giving rise to the genealogical lines that are acted on by the environment to cause deviation from the norm. Variation here is equated with deviation. There is no link back between variation and the creation of species. The development of acquired characteristics can be incorporated in such theories only by means of radical internal inconsistency.

The fixity of these natural categories is the basis for moral judgments as well. Genesis and the chain of being not only show that creation is a one-time, orderly act; they also argue that the levels of existence are ranked, with humans at the pinnacle. Human relations with nonhuman nature and with our own biological existence are orchestrated by the hierarchy of the original creation. There is no historical dynamic, no change, no possibility whatever of evolution.

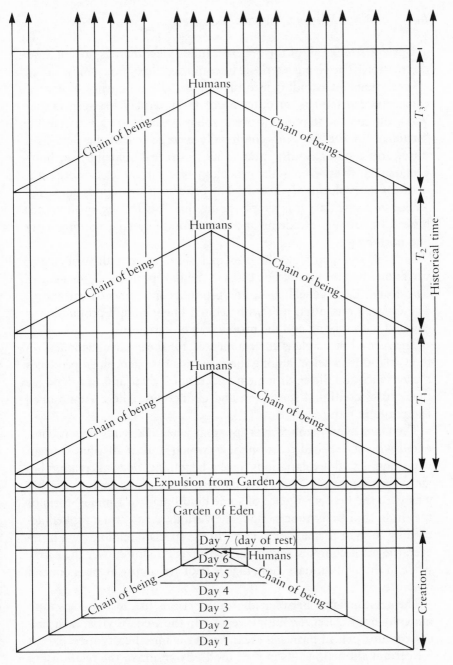

FIGURE 3   The chain-of-being view of the history of life

63

Darwin attacks all of this scheme. In his view there is no single act of creation. The living world is in continual flux. Everything varies, all variations intergrading into each other. The emergence of natural categories (species) is an outcome of the interplay between chance variation and the selection processes that act on that variation in the "struggle for life." Species continually emerge out of the flux, as it were, and eventually disappear. Their historical contribution, however, is real. Every step in the genealogy of a current species bears the mark of its evolutionary history. The process as a whole is going nowhere in particular; it is simply going on. This history of life is the history of coming into being and extinction of species, nothing more and nothing less.

In the Darwinian view, the principal observation is the primacy of variation and speciation. For Darwin the first principle is the genealogical one. Each genealogical line produces spontaneous variations. These variations often mix with related lines. Here reproduction is equated with mixing and the production of variation.

The second principle is environmental, for the environment acts on the variations to select among them. In this case genealogy produces the variance and the environment selects among variants in a dialogue that causes continual speciation and extinctions. There are no fixed categories in nature.

The Darwinian revolution is apparent when Darwin's view is compared with the special creation view (which is after all what Darwin was doing). Both explanations have a beginning. Special creation begins when God said, "Let there be light"; the evolutionary view begins with the first emergence of some unicellular (or simpler) form of life. In both, in any time period, the various taxa can be ranked according to their formal complexity. In the special-creation view, however, the most complex taxon, humans, has dominion over the earth, while in the evolutionary view, higher taxa are simply higher taxa and no more.

The relationships between the time periods in the two views are quite different. After the initial creation in the Genesis view, each successive time period basically reexpresses the initial categories, except for certain extinctions (as in the Flood). By contrast, the relationship between time periods in the evolutionary view is a genealogical relationship in which current taxa develop out of previous ones. Some are extinguished and the rest become gradually transformed by the evolutionary process.

The two views place morality in very different contexts. The doctrine that underlies the idea of special creation is that "what is, is good" because it was designed by God. The doctrine of evolutionism is that "what is, is." For evolutionists, moral imperatives do not arise from the contemplation of natural categories that God created once and for all. Moral imperatives must be drawn from elsewhere; biology cannot be the handmaiden of ethics.

The environment plays a radically different role in the two views. In the special-creation view, the deity precluded the possibility of speciation. Differing environments work on fixed species to create variance in each generation's expression of the ideal category. In the evolutionary view, by contrast, variance is continuously produced by nature itself. The action of the environment on the variance actually produces the species, the natural categories.

For the special creationists, only the categories given by God's design are real. Each generation simply reincarnates them. For Darwin, only the facts of variance and selection are real (in the sense of being constant features of life on the planet). Species are simply the transitory embodiments of the results of this opportunistic, statistical process.

Even the observer's point of view is not absolute. Darwin argued that all perception is relative to the material structures that do the perceiving. Thus even our knowledge of evolution is conditioned by the structure of our own perceptual apparatus, itself a product of evolution. The special creationists, by contrast, must believe that perception of the absolutes in the design that lies behind the world of variable appearances is possible and that the human mind is capable of it.

Thus evolutionism not only demotes humans to kinship with the other animals but also eliminates our absolute godlike perceptual distance with regard to the everyday world of existence. The "facts" of observation and history are all we can deal with; our greatest hope is to understand where we come from. In place of a world of bounded natural categories and clear moral interpretations, evolutionism places only awe and curiosity.

The Minimum Standards for an Evolutionary View of Living Things

Darwin's views created an enormous stir in the scientific community, and many subsequent thinkers have proclaimed themselves to be ardent evolutionists. But not all evolutionists agree about the implications of Darwin's ideas, and some patently non-Darwinian ideas have found their way into the biological and social sciences, as Mayr has pointed out (1982). For this reason, it is necessary to set certain minimum standards that a theory must meet if it is to be considered evolutionary.

In order to be evolutionary, a theory must embody the following propositions:

1. Variation is a ubiquitous feature of all living things. It is continually and normally produced spontaneously.

2. Selection is the result of the interaction of specific sets of environmental conditions with the variations in species of plants and animals. Selection is the force that gives rise to and alters the categories of living things.

3. The interaction between variation and selection results in adaptation or extinction. Adaptation is always relative to particular organisms and specific environments. Adaptation is never permanent.

4. All forms of life are ultimately related to each other by genealogical connections.

5. There are no nonmaterial forces at work in the evolutionary process, nor are there any "pull" factors in evolution.

6. There is no radical dichotomy between humans and other animals (between "culture" and "nature"), just as there are no radical dichotomies between any things in nature at all. Species are ranges of variation that intergrade into each other at the margins.

Though Darwin's works can be used to set standards for evolutionary thinking, they are by no means free of problems. Darwin lacked knowledge of the mechanisms of heredity and concocted the theory of pangenesis to support his argument. His stress on the absolute, slow uniformity of the pace of the evolutionary process was probably too extreme.

His views about society turned out to be incautious, though they were by no means outrageous in comparison with those of many of his popular contemporaries. His view of society as a natural and fair competitive arena and his thoughts on male dominance and eugenics now appear painfully naive and even hateful.

Of course evolutionary controversies continue. The phenotype/genotype distinction, the units of evolution—especially above the species level and below the individual level (i.e., biochemical processes)—the question of whether or not adaptation leads to optimum solutions, and the old chestnut about the direction of evolution are issues that continue to call forth heated debate. Yet what impresses one most in reading the literature produced by such contemporary evolutionary biologists as Ernst Mayr, Sewall Wright, Richard Lewontin, and Stephen Jay Gould is the degree to which Darwin's basic vision remains viable well over a century after its first public expression.

The controversies in modern evolutionary biology are important and lively. While some tenets of evolutionary thinking are agreed upon, mechanisms and interpretations are hotly debated. Exclusive attention to these contested issues in recent decades has led attention away from some larger questions about the acceptance of evolutionism, questions that only recently have begun to surface again.

The literature on racism and eugenics has been shown to contain numerous attempts to claim the support of evolutionary theory and other scientific views for a variety of oppressive doctrines (Chase 1975–76). The continuities between pre-evolutionary and current views on these issues are depressingly apparent and are well known. But other kinds of continuities exist.

Recently evolutionary biologists who approach the problem in very different ways have begun to argue that elements of pre- and non-evolutionary thinking are much more important in modern biology than current practitioners imagine. Mayr, in *The Growth of Biologi-*

cal Thought (1982), distinguishes between "essentialist" thinking (what I have described as humoral/environmental theories) and "population" thinking (evolutionary reasoning). Mayr contends that the rise of evolutionism involves the replacement of essentialism by population thinking. The conflict between these two views, however, is of long standing in the Western intellectual tradition, and it continues to plague modern evolutionary biology. Though Mayr's agenda in contrasting these two traditions is somewhat different from mine, his view of the history of biology closely supports my characterization of the differences between pre- and nonevolutionary views of nature and evolutionary ones.

Another major trend in recent biological thought is the critique of "adaptationism." This critique has not been connected to the historical and interpretive issues I am emphasizing, but I am convinced that it can and should be. The adaptationist critique is most closely associated with Gould and Lewontin, who criticize certain contemporary biologists for introducing an adaptationist view into their evolutionary thinking:

> . . . the adaptationist programme, or the Panglossian paradigm . . . is rooted in a notion [of] the near omnipotence of natural selection in forging organic design and fashioning the best among possible worlds. This programme regards natural selection as so powerful and the constraints upon it so few that direct production of adaptation through its operation becomes the primary cause of nearly all organic form, function, and behaviour. . . . An organism is atomized into 'traits' and these traits are explained as structures optimally designed by natural selection for their functions. . . . After the failure of part-by-part optimization, interaction is acknowledged via the dictum that an organism cannot optimize each part without imposing expenses on the others. . . .
>
> The adaptationist programme can be traced through common styles of argument . . .
>
> (1) If one adaptive argument fails, try another. . . .
> (2) If one adaptive argument fails, assume that another must exist. . . .
> (3) In the absence of a good adaptive argument in the first place, attribute failure to imperfect understanding of where an organism lives and what it does. . . .
> (4) Emphasize immediate utility and exclude other attributes of form. . . . [Gould and Lewontin 1979:584–87]

From these premises Gould and Lewontin go on to argue that these fallacies in action result in the telling of "adaptive stories" that are immune to test.

Gould and Lewontin are combating a form of biological reasoning that reproduces important elements of the pre-evolutionary design argument. Perfect adaptation as an assumption belongs nowhere in evolutionary biology. It does, however, fit into the concepts of nature found in the design view, the special-creation theory, and some applications of humoral/environmental theory. Fixed species, clear natural categories, and perfect adaptations are the antithesis of evolution.

To use biological science to tell stories about a world in which all the organic parts are at an adaptive optimum is in effect not to use biological science at all, but to reintroduce teleology and theology into the study of the organic world. It is quite consistent with general attempts to domesticate Darwinism's randomized, liminal world in motion and render it less fearsome.

Together the works of Mayr, Gould, and Lewontin show that even within modern evolutionary biology, significant strains exist. The conflict between evolutionary and nonevolutionary views of nature has not come to an end with the "modern synthesis."

II *Simple Continuities*

Pre- and nonevolutionary views have persisted to the present in two basically different ways. Though they are not always easy to distinguish at the margins, they require separate treatment. The views that may be called simple continuities are the subject of Chapter 3. Complex continuities are given extended treatment in Part III.

Simple continuities are the direct persistence of pre-evolutionary views in the work of post-Darwinian scholars. Despite a slim overlay of evolutionary terms, many of these works simply reproduce the fundamental theoretical structures of pre-evolutionary thinking.

Though simple, these continuities are important socially and intellectually, as they have played a large role in some of the darker moments of modern racism and eugenics. They are also rather well documented as a result of the great attention devoted to the history of racist thought. I therefore treat them only briefly.

CHAPTER 3

Humoral Politics:
Races, Constitutional Types, and
Ethnic and National Characters

Nowhere is the absence of a full-scale Darwinian revolution more apparent than in the literature on race, ethnic and national characters, and constitutional types. Post-Darwinian works in this area are a direct continuation of the pre-Darwinian tradition. With the exception of some stray evolutionary terminology, many contemporary works are indistinguishable from their ancestors. This is a case of descent without modification.

In sheer bulk, Western ideas about the differences between races, nations, and constitutional types and their relationship to differing environments have probably constituted the dominant use of humoral/environmental ideas. The number of tracts written on these subjects over the span of European history defies cataloguing. Nor are these the concerns of some restricted class of intellectuals. We know from both current experience and the historical record that all classes of people from all walks of life use racial, ethnic, and national character ideas and constitutional typologies to explain the ways they behave, the ways other people behave, and why some groups are seemingly inferior or superior to others. Anthropologists place these ideas at the very center of the Western explanations of group similarities and differences.

The Hippocratic Tradition

Toward the end of his discourse on airs, waters, and places, Hippocrates explains the differences between racial groups as the result of humoral/environmental factors. Comparing Asia and Europe, he stresses how little they resemble each other physically. He then says:

> ... Asia differs very much from Europe as to the nature of all things, both with regard to the productions of the earth and the inhabitants, for everything is produced much more beautiful and large in Asia; the country is milder, and the dispositions of the inhabitants also are more gentle and affectionate. The cause of this is the temperature of the seasons, because it lies in the middle of the risings of the sun towards the east, and removed from the cold (and heat), for nothing tends to growth and mildness so much as when the climate has no predominant quality, but a general equality of temperature prevails ... the inhabitants too, are well fed, most beautiful in shape, of large stature, and differ little from one another either as to figure or size; and the country itself, both as regards its constitution and mildness of the seasons, may be said to bear a close resemblance to the spring. Manly courage, endurance of suffering, laborious enterprise, and high spirit, could not be produced in such a state of things either among the native inhabitants or those of a different country, for there pleasure necessarily reigns. [Hippocrates 1886:169–70]

In his often cited discussion of the Macrocephali he stressed the physical differences that distinguish these people from other races. The cause is environmental, now in a special sense. The Macrocephali deformed their heads environmentally (that is, by custom), but eventually the deformity came to be inherited. In this view, culture is an effective force in shaping human bodies because it operates by physical means to alter the humoral balance.

This idea introduces an ambiguity into his explanatory strategy. Environmental factors may bring about physical changes in humans either independently or through the effects of cultural practices. The problem, then, is to know when the cause of a physical characteristic is cultural and when it is not. In the case of the Macrocephali, he feels confident of the cause because their head shape is such an oddity.

But how could this approach help one to distinguish between the "natural frame" of a population and its cultural alterations? Logically the only way to make such a distinction would be to move people into different environments or to change their cultural practices and then see what aspects of their physical and behavioral makeup did not change. But such an experiment is effectively impossible to perform, just as contemporary attempts to segregate heredity and environment are compromised both theoretically and methodologically (Lewontin [1974] 1976). Stereotypes about different populations and environments carry most of the explanatory weight.

This is an important part of the conceptual foundation of Western thought about races, ethnic groups, nations, and constitutional types. To this day it seems common-sensical to explain the character of people to some extent in terms of the place they have come from. Racial slurs, and racial praise as well, are often linked to place—the interiorization of the external environment is widely used as a principle for explaining human behavior. In the Western world, at least, the question "Where are you from?" really means "What kind of person are you?" That most of us know where we come from at least a few generations back and where most of our friends and colleagues are from tells us something about the continuing power of the connection between people and places in our systems of classification.

The ambiguities in Hippocrates' views are also still with us. The problem of distinguishing between environmental influences caused by nonhuman and human agents is compounded by the tendency of people to move around. Once people have moved away from their place of "origin," the theory encounters a serious problem.

The basis of the theory is a direct, immediate material connection between the environment, the humors, and behavior. When a group of people who have lived for many generations in one kind of environment move to another, theoretical consistency demands that the new environment should take over completely and modify the humoral constitution of the population. But this idea conflicts both with a theoretical principle in Hippocrates' text and with common-sense observation.

The genealogical principle in humoral/environmental theory insists that people get their humoral constitutions not merely from the direct action of the environment but from their parents by physical transmission. The observation of parent-child similarities and the concept of blood as the source of shared physical substance among kin suggests that people are what they are in part because of the people they are descended from, and not merely because of where they are. A groups that moves into a new environment will still be partly formed by its genealogical connections with its past and will only gradually change in the new environment, possibly by means of the sorts of mechanisms Hippocrates discusses in regard to the Macrocephali.

Thus the total application of the humoral/environmental theory as a direct explanation of the character of peoples through immediate physical causes seems very powerful, but too strict an application of it comes into conflict with the genealogical principle of humoral constitution and with the observed persistence of cultural practices. Thus an enormous breach is opened in the conceptual structure. Any behavior not immediately attributable to local humoral/environmental causes can be attributed to genealogical connections by way of inherited humors from other environments or as survivals. Any failure of people to behave as their families, ethnic groups, or national governments wish can be met with the view that their excellent genealogy has been sullied by the nefarious influence of a hostile environment or mixture with "inferior" races.

It is apparent that this breach has provided the basis for most of the arguments about nature versus nurture right up to the present. The extreme environmentalist explains deviations from expectations by invoking genealogical and/or cultural causes that prevent the environmental influences from fully expressing themselves. The extreme genealogist (and likewise the cultural determinist) can explain deviations from the theoretically expected behaviors by invoking the contaminating influences of environment.

Jean Bodin (1529[?]–1596)

In the seminal works of Jean Bodin (heavily relied upon by Montesquieu and many others) we can see a clear continuity between the classical theories of the humors and environments on the one hand and contemporary political theory on the other. Bodin lived during the great wars of religion and the major European overseas expansions. In his major works, *Six Books of the Commonwealth* and *Method for the Easy Comprehension of History* (published in 1576 and 1583, respectively), he develops a theory of the political management of states based on an analysis of the influence of the environment on human behavior. He attempts to differentiate what is "natural," and therefore beyond human control, from what is "political," and thus subject to human design. He examines cultural differences around the world and in doing so provides exemplifications of the ambiguities in the humoral/environmental theory which are still with us.

In Book 5 of *Six Books of the Commonwealth*, Bodin postulates that the diversity of races requires different kinds of commonwealths.

Political institutions must be adapted to environment, and human laws to natural laws. Those who have failed to do this, and have tried to make nature obey their laws, have brought disorder, and even ruin, on great states. One observes very great differences in the species of animals proper to different regions, and even noticeable variations in animals of the same species. Similarly, *there are as many types of men as there are distinct localities.* Under the same climatic conditions oriental types are different from occidental, and in latitudes at equal distances from the equator, the people of the northern hemisphere are different from those of the south. What is more, when the climate, latitude, and longitude [are] the same, one can observe variations between those who are mountaineers, and those who live on the open plains. Even in the same city there is a difference in humour and in habits between those who live in the upper and those who live in the lower parts of the town. This is why cities built in hilly country are more subject to disorders and revolutions than those situated on level ground. ... *A wise ruler of any people must therefore have a thorough understanding of their disposition and natural inclinations before he attempts any change in the constitution or the laws.* One of the greatest, if not the principal, foundation of the commonwealth is the suitability of its government to the nature of the people, and of its laws and ordinances to the requirements of time, place, and persons. For although Baldus says that reason and natural equity are not conditioned by time and place, *one must distinguish between universal principles, and those particular adaptations that differences of places and persons require.* The governments of commonwealths must be diversified according to the diversities of their situations. The ruler must emulate the good architect who builds with the materials locally available. The wise statesman must do this too, for he cannot choose such subjects as he would wish. [Bodin (1576) 1955:145–46; emphases mine]

In developing a set of associations between particular environments and particular kinds of human beings, Bodin elaborates on geopolitical stereotypes that are with us to this day. "Northerners succeed by means of force, southerners by means of finesse, people of the middle regions by a measure of both" (p. 148). He links these characteristics to the humors:

Those who live at the extremities near the poles are phlegmatic and those in the extreme south, melancholic. Those who live thirty degrees below the pole are of a more sanguine complexion, and those who are about midway, sanguine or choleric. Further south they become more choleric or melancholic. They are moreover tanned black or yellow, which are the colours of black melancholy and yellow choler. [Pp. 149–50]

Note that the humors are clearly the source of our contemporary color classification of races: white/phlegm, yellow/yellow bile, red/blood, and black/black bile. This connection between humors and races must be widely known but I have not found a reference to it in histories of racism.

The humoral color itself is insignificant, as people do not really come in these colors. Rather the color associates the predominance of a humor with a behavioral stereotype associated with each racial category. Thus American Indians are savage and warlike; blacks are lethargic and slow-witted; Orientals are cunning; and whites, not surprisingly, are reflective and rational.

Bodin equates the northern, southern, and temperate zones with the three ages of man: youth, old age, and maturity. "Northerners rely on force, those in the middle regions on justice, and southerners on religion" (p. 151). Predictably he decides that the people in the temperate zone are best suited for managing commonwealths.

Throughout his two major books, Bodin is concerned to understand the "natural" inclinations of different races in order to do two things. First, he wants to evaluate their historical deeds fairly. He believes it is unfair to criticize or to praise a northerner for being warlike, since belligerence is his "natural inclination." Correct evaluation of history depends on understanding what people are constrained to be like, not blaming or crediting them from behaving as they naturally must.

Second, as a political reformer, Bodin wants to be able to suggest the form of government that would fit the "natural inclinations" of each race of people, since it is impossible to govern all people as if they were the same. The humoral/environmental differences among them would make such a project fail. Bodin demands a recognition of natural diversity harnessed to a program of political diversity in the name of a more abstract general human equality.

Bodin himself is clearly troubled by two parts of his own argument, and both are relevant to our concerns. Characteristically for his time, he worries about the religious implications of his views. According to the literal interpretation of the Bible, God created all humanity. How, then, could God have created humans of unequal abilities? Bodin provides a very weak answer:

> Not that God respects either places or peoples, or fails to put out His divine light over all. But just as the sun is reflected more brilliantly in clear still water than in rough water or a muddy pool [read northern and southern races], so the divine spirit, so it seems to me, illumines much more clearly pure and untroubled minds than those which are clouded and troubled by earthly affections. [Pp. 152–53]

The argument makes little sense. If God is all-powerful, then creation is not constrained by rough waters or muddy pools. Ultimately his views assert the inequality of races. His view of inequality, connected with the political argument that matches different kinds of people to different forms of government, justifies both the suppression of minorities and imperialism. It is a short step from Bodin's position to the polygenist view of races as separate creations.

Another problem Bodin faces is the relationship between the "natural" inclinations of a people and the state's ability to modify these inclinations by cultural means. The conflict between nature and nurture is dramatically expressed: "Whatever laws or ordinances are made to diminish it, the natural inclination of the people will always reassert itself" (p. 152). Yet "there are of course men of all kinds of temperament in all localities and countries, though more or less subject to these general conditions which I have described. Moreover the particular can greatly modify the general character of the country" (p. 154).

Then, moving even farther in this direction, Bodin says:

> If anyone would understand how nurture, laws, and customs have power to modify the natural disposition of a people, he has only to look at the example of Germany. In Tacitus' day its inhabitants knew neither laws, religion, the sciences, nor any form of commonwealth. Now they are second to none in all these achievements. . . . On the other hand the Romans have lost the greatness and virtue of their fathers and are nowadays idle, mean, and cowardly. . . . *If the discipline of laws and customs is not maintained, a people will quickly revert to its natural type.* [Pp. 156–57; emphasis mine]

Finally, he proposes to derive policy guidelines from this complex web of conflicting causes:

> So much for the natural inclinations of peoples. As I have said, this compulsion is not of the order of necessity. But it is a very important matter for all those who are concerned with the establishment of the commonwealth, its laws and its customs. They must know when and how to overcome, and when and how to humour these inclinations. [P. 157]

Thus people are as they are because of the place their race came from originally. They bear the humoral/environmental stamp of the locality. At the same time, culture has a capacity to modify these natural inclinations and to bring great changes in people's behavior. But Bodin also says that when political discipline is relaxed, people always have a tendency to revert to their "natural type" (which presumably is maintained through the genealogical connections with the environments in which they were formed).

Thus in Bodin's view a kind of internal war is waged between nature and culture, and politics is enlisted in the battle on culture's side. Humans are dominated by direct environmental influences, but we are also genealogically connected to past environments and culturally able to modify our behavior against the environment's demands. These contradictions are politically important because Bodin justifies political action as the result of an impartial assessment of the "natural types" of people in different places.

Since there is no impartial way of deciding what any person's "natural type" is, how do we know when people are acting against their "natural type"? Only by means of agreed-upon stereotypic national, ethnic, or racial histories. Thus Bodin's appeal to nature as a guide to political conduct leaves an open field for manipulation while justifying ethically any of a host of political interventions.

Constitutional Types

A more specific connection between environments and physical types is made in the large literature on constitutional types. A tremendous diversity of material is available on the relationship between the physical constitutions of people, the environment, and their behavior. There are even putative correlations between certain physical types and position in the social hierarchy.

Though this intellectual arena has not been treated integrally by scholars, it is certainly well organized by people in the ordinary business of life. People respond differently to people of differing physiotypes. Leaving aside sexual differences, there are clearly somewhat different expectations for tall, thin people than for short, fat ones, for dark people than for light people, and so on. When people are asked to make up stories about individuals whose appearances differ, the stories vary in accordance with the appearances described. These complex symbologies of the body and its adornment have not yet been given the attention they require.[1]

In multicultural situations, differences in aesthetic preferences and in the general symbology of the body often lead to alienation or at least confusion. People who are perceived to be too dark, to flap their arms too much, or to move too slowly are hard to deal with. Each era has its caricatures of the physiotypes associated with different social classes and ethnic groups. Pat and Mike are more than names; they invoke appearances as well. How many medieval representations of a usurer who was tall, well-muscled, and light are there? Apparently, then, these physiotypes play an important role in the construction of the everyday world of meanings.

Burton's *Anatomy of Melancholy*

Throughout the history of Western thought attempts have been made to systematize categories of constitutional types into more fully organized and presumably more scientific systems. Robert Burton's *Anatomy of Melancholy*, a famous example of this genre published in 1621, is now generally treated as a work of literature. Burton sought to explain and to list all of the manifestations and cures of the predominance of black bile, the humor believed to cause melancholy. He concentrated on melancholy because he considered it the most pernicious of all the humoral disorders.

The scope of afflictions included under the heading of melancholy is wide, from "inequality" to mental illness. Each ailment is carefully defined and subdivided, and a hierarchical order of treatments is set forth. In Burton's view, all behavioral states are caused by the effects of humors and the impact of the environment on them. Individuals differ in the degree to which these influences express themselves and in their responsiveness to treatments designed to counteract them.

The 1338-page modern edition of Burton's book is exceedingly rich in the detailed analysis of humoral concepts and constitutional types; it includes a multitude of references and constructs a complex world of sickness and health on the basis of humoral ideas. Between Burton and present-day thinkers stretches a long tradition of the analysis of constitutional and racial types. Phrenology and other such typological systems have always had great appeal. Concepts of the neurasthenic young romantic, the gluttonous, gout-ridden old man, and the tubercular heroine were and are widely used in both literature and medicine.[2]

Francis Galton

No clear distinction is to be made between ideas about constitutional types and popular Western views on fixed racial characters.

Already in Hippocrates and Bodin we have seen the connection between ideas about racial characters and humoral/environmental conditions. From them it is only a short step to the racial and eugenic ideas of Francis Galton.

This fact is well known. It should be regarded as a problem to be explained, however, because Francis Galton wrote after the publication of Darwin's *Origin of Species* and claimed to be a thoroughgoing evolutionist. The continuity between Galton and his pre-Darwinian predecessors is thus an example of the kind of simple continuity in humoral/environmental (and thus nonevolutionary) ideas I seek to uncover.

Galton, Darwin's cousin, is well known for his numerous writings on statistics, fingerprinting, twin studies, blood transfusions, criminology, meteorology, and travel. He is perhaps best remembered for his work in eugenics. His fame began in 1869 with *Hereditary Genius*, a book that Darwin cited with admiration. Written in the language of Darwinian evolutionism, the book contains a complex statistical apparatus that tries to make his conclusions appear to be the results of scientific investigation. Much of the argument is still current today.

Galton begins by stating that he wrote the book to contradict the idea that the mind is free of the effects of "natural laws." From his vantage point, there are only two sources of human behavior: the inborn (hence natural) and the acquired (hence cultural). It is his intention to place genius on the side of nature, just as Hippocrates and Bodin did. Galton, however, was an unabashed racist.

> The natural ability of which this book mainly treats, is such as a modern European possesses in a much greater average share than men of the lower races. There is nothing either in the history of domestic animals or in that of evolution to make us doubt that a race of sane men may be formed, who shall be as much superior mentally and morally to the modern European, as the modern European is to the lowest of the Negro races. [Galton (1869) 1962:27]

Galton saw races as fixed ideal types:

> ... differences ... between men of the same race might theoretically be treated as if they were Errors made by Nature in her attempt to

> mould individual men of the same race according to the same ideal
> pattern. Fantastic as such a notion may appear to be when it is ex-
> pressed in these bare terms . . . it can be shown to rest on a perfectly
> just basis. [P. 28; emphasis removed]

This and related arguments about regression to the filial, parental,
and racial "center" demolish variation as the source of human evolution. This idea alone severs the connection between Darwinism and
Galton's views. The denial of the evolutionary potential of variability
is crucial for Galton because it allows him to argue for a social policy
of eugenics to control the future of humanity. He argues against any
other possible sources of human change in order to justify his policy.

How, then, did the races get their characteristics? Galton is ambiguous on this point. He speaks of the "natural refinement" of the Hu-
guenots (p. 38) and says that "the natural temperaments and moral
ideas of different races are various" (p. 39). It appears that races are
as they are because they were as they were.

Later the argument becomes even less clear. He calls the British a
race; judges a race; lowland Scots, English northcountrymen, and
Athenians races. He also speaks of "civilized races." In a peculiar
passage, he even cites the "race-destroying" influences of "heiress
blood." It seems that men of great ability tend to marry heiresses, and
as heiresses tend to come from relatively infertile families (or the women
would not inherit large estates), such matches lead to low fertility in
the families of men of great ability.

Given the array of uses of the term "race," it is hard to pin down
what Galton sees as the cause of racial differences. It is clear that he
believes that races became as they are long ago. He sees the effect of
civilization as "either . . . modifying the nature of races through the
process of natural selection whenever the changes were sufficiently
slow and the race sufficiently pliant, or of destroying them altogether
when the changes were too abrupt or the race unyielding" (p. 399).
And in passages that sound remarkably like arguments of Konrad
Lorenz ([1963] 1971), Robert Ardrey (1966), and Desmond Morris
(1969), Galton argues that there is a lack of synchrony between the
requirements of modern civilization and the habits we have derived
from our savage ancestors. The moral of the story is that we must
intervene to make ourselves suitable to our civilization.

In Galton's view, a race cannot be much improved by natural selection, but it can be ruined. When the original "natural" races (which were apparently not created equal) are intermixed, the fine qualities of the naturally higher races are corrupted. Indeed, he argues that the degeneration of Athens was caused by the loss of its racial "purity."

The core chapters of the work are surveys of various categories of preeminent people in Britain. Galton argues that superior men are "naturally" better equipped to compete in society than are others. They successfully overcome social hindrances and rise to the top quickly. To support this proposition, he attempts to trace the statistical incidence of genius in the higher occupational groups. Not surprisingly, he finds that hereditary genius makes men successful.

The serious social problem he sees is that these superior men show a markedly lower fertility than do inferior men. Society will be ruined unless we take control of our collective destiny and develop a positive eugenics program. He closes with the following statements:

> Nature teems with latent life, which man has large powers of evoking under the forms and to the extent which he desires. We must not permit ourselves to consider each human or other personality as something supernaturally added to the stock of nature, but rather as a segregation of what already existed, under a new shape, and as a regular consequence of previous conditions. . . . There is decidedly a solidarity as well as a separateness in all human, and probably in all lives whatsoever; and this consideration goes far . . . to establish an opinion that the constitution of the living Universe is a pure theism, and that its form of activity is what may be described as co-operative. . . . It also suggests that they [individuals] may contribute, more or less unconsciously, to the manifestation of a far higher life than our own, somewhat as—I do not propose to push the metaphor too far—the individual cells of one of the more complex animals contribute to the manifestation of its higher order of personality. [Pp. 427–28]

This sort of comprehensive justification for eugenic controls and the suppression of individual freedoms has been made often since Galton.

Galton clearly regarded nature as a realm of physical laws, in contrast with culture, which is a realm of rational argument and action. He regularly contrasts law with will, ability with education, instinct with reason, and fixed with variable phenomena. While all these contrasts are important, the last one is startling. To refer to hereditary

phenomena as fixed is to contradict a core principle of the evolutionary doctrines Galton claims to be relying on.

Galton's argument hangs on the fixity of species; his only dynamic
idea is that races degenerate through intermixture. The "natural" races
are fixed unless we use culture to improve them. If we do not control
nature with our reason, then the continuing mixture of races will ultimately breed away all outstanding human characteristics.

The book is peppered with references to purity of blood, mixture
of blood, breeds, and other notions that show that Galton's views on
social hierarchy and racial differences were drawn from pre-evolutionary ideas. Galton believed that people are socially superior because of their superior bloodlines. It thus appears that Galton simply
updated part of the terminology of the humoral justification for social
inequality. The more recent history of such racist thought has been
admirably analyzed by Stephen Jay Gould in *The Mismeasure of Man*
(1981).

Ernst Kretschmer

Galton's celebrity makes him a frequent target for just the sort of
criticism I am offering. But the simple continuation of pre- and nonevolutionary ideas about "human nature" occurs in many less spectacular contexts. It remains influential and touches many lives in important ways.

The psychiatrist Ernst Kretschmer endeavored to use constitutional
typology to place the diagnosis and treatment of mental illness on a
scientific basis. He began his major work, *Physique and Character*
(first published in English in 1921), with the following statement:
"Investigation into the build of the body must be made an exact branch
of medical science. For it is one of the master-keys to the problem of
the constitution—that is to say, to the fundamental question of medical and psychiatric and clinical work" (Kretschmer 1925:5). Although body form is not identical with the constitution, the constitution of an individual can be inferred from his or her body form.
Kretschmer felt the major problem in getting at this form was the lack
of a refined typological system. He thus created the following scheme
for diagnostic observation (the subcategories have been omitted):

CONSTITUTION SCHEME

I. FACE AND SKULL

Head, Eyes, Nose, Mouth, Lips, Cheekbone, Lower Jaw, Chin Larynx, Teeth, Gums, Ears, Forehead, Profile, Frontal, Outline, Facial form, Cranium, Back of Head

II. PHYSIQUE

Poise, Bone structure, Joints, Musculature, Fat upholstery, Head, Legs, Hands, Feet, Shoulders, Chest, Stomach, Spine, Pelvis

III. SURFACE OF THE BODY

Skin, Blood-vessels, Hair

IV. GLANDS AND INTESTINES

Testicles, Genitalia, Thyroid gland, Lymph-glands, Mammary-glands, Internal diseases

V. MEASUREMENT

Height, Circumference, Length, Breadth, Skull

VI. TEMPORAL

Commencement of mental disturbance, Commencement of puberty, Commencement of involution, Commencement of fattening, Commencement of emaciation, Commencement of baldness, Commencement of certain physical disorders, Sexual abnormalities

VII. SUMMING UP OF PHYSICAL STATES

VIII. TYPE OF PERSONALITY

IX. HEREDITY

[Pp. 5–8]

Each of these general categories has a set of subcategories for measurement and classification. Despite its apparently large number of categories, this scheme is exceedingly simple. Anyone with a little experience in anatomical description would find it adequate only for getting at the grossest features. It is also static: the body in motion cannot be captured in this typological system. Yet Kretschmer is reasonably satisfied that this scheme can be used to put constitutional typology on firm scientific ground.

To "test" the classification and to provide a set of statistical values for comparison, Kretschmer applied this scheme to a population of

Schwabians. He enters an interesting caveat: ". . . without further work they [the results] cannot be compared with numerical values obtained from patients belonging to *other races*" (pp. 12–13; emphasis mine).

Though he is quite properly cautious about the problem of statistical sampling, he simply calls the Schwabians a "race." While this usage was commonplace in references to many regional populations in Europe and the United States, it suggests a fundamental primitivism in Kretschmer's grasp of the methodological and theoretical problems of biological taxonomy, despite his belief that he was making a major contribution to just this field.

Kretschmer is stern in his criticism of previous attempts at constitutional typology, particularly their lack of objectivity. To replace previous classificatory types, he introduces his own general types: the asthenic, the pyknic, and the athletic, and a mongrel misfit or antitype called the dysplastic. Figures 4, 5, and 6 show the photographs he offers as examples of the first three types. Armed with these types and a set of arguments about the behavioral tendencies associated with each, he constructs his theory of temperaments.

He believes these types to be "real" in some ultimate sense: "The types . . . are not 'ideal types' which have emerged . . . in accordance with any given guiding principle of collection or pre-established values. They are . . . obtained from empirical sources . . ." (p. 18). And he says:

> It is the same here as in clinical medicine, or in botany or zoology. The 'classic' cases, *almost free from any mixture*, and endowed with all the essential characteristics of *a perfect example of some form of disease, or a zoological race-type*, are more or less lucky finds, which we cannot produce every day. From this it follows, that *our description of types . . . refers not to the most frequent cases, but to ideal cases . . .* [P. 19; emphasis mine]

This strange bit of illogic is singled out to show that Kretschmer, despite claims of objectivity, is actually working with a small set of aesthetic ideals drawn from Western culture, ideals of which real people are more or less (usually less) perfect embodiments.

He characterizes the physiotypes in the following manner. His asthenic type is described as having "a deficiency in thickness combined with an average unlessened length." The asthenic type is "a lean narrowly-built man, who looks taller than he is, with a skin poor in secretion and blood, with narrow shoulders, from which hang lean arms with thin muscles" (p. 21; emphasis removed). His pyknic type

> is characterized by the pronounced peripheral development of the body cavities ... and a tendency to a distribution of fat about the trunk, with a more graceful construction of the motor apparatus ... middle height, rounded figure, a soft broad face on a short massive neck, sitting between the shoulders; the magnificent fat paunch protrudes from the deep vaulted chest which broadens out towards the lower part of the body. [P. 29; emphasis removed]

The male athletic type

> is recognised by the *strong* deveopment of the skeleton, the musculature and also the skin. . . . A middle-sized to tall man, with particularly wide projecting shoulders, a *superb* chest, a *firm* stomach, and a trunk which tapers in its lower region, so that the pelvis, and the *magnificent* legs, sometimes seem almost *graceful* compared with the size of the upper limbs and particularly the hypertrophied shoulders. . . . The outlines and shadings of the body are determined by the swelling of the muscles of the good or hypertrophied musculature which stands out plastically as muscle-relief. [P. 24; emphases mine]

There can be little question which type is the best.

Kretschmer correlates these types with "circular" (manic-depressive) and "schizophrene" psychic dispositions (Table 1). From these data he concludes that

> we can formulate our results straight away. (1) There is a clear biological affinity between the psychic disposition of the manic-depressives and the pyknic body type. (2) There is a clear biological affinity between the psychic disposition of the schizophrenes and the bodily disposition characteristic of the asthenics, athletics, and certain dysplastics. (3) And vice versa, there is only a weak affinity between schizophrene and pyknic on the one hand, and between circulars and asthenics, athletics, and dysplastics on the other. [P. 36]

A long set of chapters on various forms of mental illness in relation to constitutional types is followed by a chapter on the relation be-

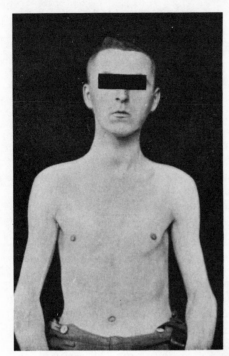

FIGURE 4   Kretschmer's asthenic type, described as "schizoid psychopath" (Kretschmer 1925, pl. 1

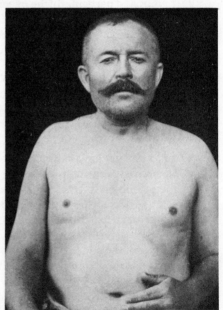

FIGURE 5   Kretschmer's pyknic type, described as "circular" (manic-depressive) (Kretschmer 1925, pl. 3)

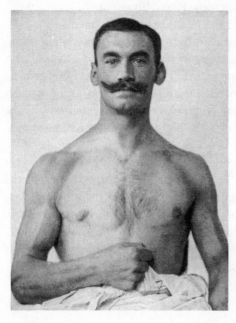

FIGURE 6 Kretschmer's athletic
type, identified with "schizophrenia"
(Kretschmer 1925, pl. 3)

TABLE 1

Number of Kretschmer's subjects who displayed "circular" and "schizophrene" psychic dispositions, by body form

Body form	Psychic disposition	
	Circular[a]	Schizophrene
Asthenic	4	81
Athletic	3	31
Asthenico-athletic mixed	2	11
Pyknic	58	2
Pyknic mixture	14	3
Dysplastic	—	34
Deformed and uncatalogable forms	4	13
Total	85	175

[a]Manic-depressive.
SOURCE: Kretschmer 1925:35.

tween constitutional type and genius. Kretschmer concludes the work with an attempt to formulate a theory of the temperaments.

Temperament, Kretschmer says, is "co-determinate with the *chemistry of the blood, and the humours of the body.* Their physical correlate is the brain-glandular apparatus. *The temperament is that class of mental events which is correlated with the physique, and probably through the secretions*" (p. 252; emphases mine). He then groups the temperaments into two major classes, cyclothymes and schizothymes, which he sums up in Table 2.

I submit that Kretschmer's entire formulation was not based on his observations at all, but on a resystematization of the theory of the humors. He uses humoral language, and we can see in Table 2 a close connection to the older humoral classifications. From top to bottom we move from those characterized by behaviors typical of the prevalence of blood and yellow bile down through the phlegmatic to the melancholic. Figures 4, 5, and 6 fit this classification equally well.

Virtually nothing of evolutionary biology is found in Kretschmer's

TABLE 2
Psychological and physical characteristics associated with Kretschmer's two classes of temperament

	Cyclothymes	*Schizothymes*
Psychaesthesia and mood	Diathetic proportion: between raised (gay) and depressed (sad)	Psychaesthetic proportion: between hyperaesthetic (sensitive) and anaesthetic (cold)
Psychic tempo	Wavy temperamental curve: between mobile and comfortable	Jerky temperamental curve: between unstable and tenacious alternation mode of thought and feeling
Psychomotility	Adequate to stimulus, rounded, natural, smooth	Often inadequate to stimulus: restrained, lamed, inhibited, stiff, etc.
Physical affinities	Pyknic	Asthenic, athletic, dysplastic, and their mixtures

SOURCE: Kretschmer 1925:258.

theory; his whole system can be explained simply as an application of humoral doctrines. His approach is typological, resting on the assumption that natural categories and ideal types actually exist in nature. Variation is treated as deviation from the ideal type, and ideal types are seen as static.

W. H. Sheldon

The major constitutional typologist of recent times is W. H. Sheldon, whose most widely read book, *Varieties of Human Physique*, was published in 1940. Sheldon uses a much more complex classificatory scheme than Kretschmer did, and he bases it on embryology. Constitutional features that show a predominance of tissue derived from the endodermal embryonic layer are called "endomorphic," and persons with a predominance of these characteristics are called "endomorphs." He describes them as having a "relative predominance of soft roundness through the various regions of the body. When endomorphy is dominant the digestive viscera are massive and tend relatively to dominate in the bodily economy" (Sheldon 1940:5).

When tissues derived from the mesoderm are predominant, the type is called "mesomorph." This physique has a "relative predominance of muscle, bone, and connective tissue. The mesomorphic physique is normally heavy, hard, and rectangular in outline . . ." (p. 5). Finally, "ectomorphy" refers to the predominance of tissues derived from the ectoderm and thus of "linearity and fragility. In proportion to his mass, the ectomorph has the greatest surface area and hence relatively the greatest sensory exposure to the outside world" (p. 5).

Starting with these three general types, Sheldon and his team of researchers took some 4,000 photos (see Figure 7) and rank-ordered them according to the predominance of each component in five regions of the body for a total of fifteen rank orders. Then anthropometric measurements made of the photographs eventually resulted in eighteen anthropometric indices. Using all these measurements, they eventually came up with a three-digit numeral to characterize each photograph. By this process Sheldon hoped to create an objective science of the relationship between temperament and constitutional type.

Sheldon asserts that these body types correlate with temperaments in the following way:

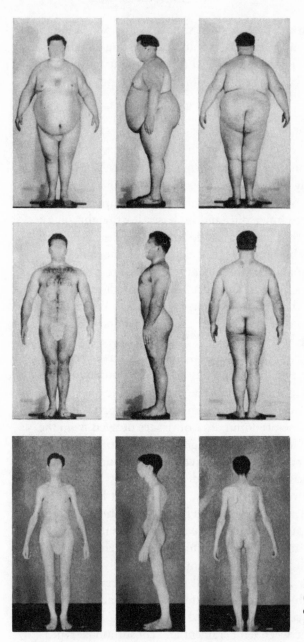

FIGURE 7   Sheldon's endomorph (*top*), mesomorph (*middle*), and ectomorph (*bottom*) (Sheldon 1940, frontispiece)

Basic aspects of temperament have been identified, objectified by the method of tests and interviews, and scaled on 7-point scales. These components we refer to as *viscerotonia, somatotonia,* and *cerebrotonia.*

Viscerotonia is roughly identifiable with love of comfort, relaxation, sociability, conviviality, and sometimes with gluttony. It is the motivational organization dominated by the gut and by the function of anabolism. Somatotonia is the motivational pattern dominated by the will to exertion, exercise and vigorous self-expression. It is the drive toward dominance of the functions of the *soma.* Cerebrotonia refers to the attentional and inhibitory aspect of temperament. In the economy of the cerebrotonic individual the sensory and central nervous systems appear to play dominant roles. He is tense, hyperattentional and under strong inhibitory control. His tendency is toward symbolic expression rather than direct action.

These components of temperament appear to correlate with patterns of somatotypes, and like the morphological components, they combine in various proportions in different individuals. They behave, within limits, as independent variables. [Pp. 8–9; emphasis his]

These somatotypes and their associated mental orientations are compatible with Kretschmer's constitutional types and with earlier humoral typologies. Sheldon multiplied the measurement techniques but he did not change the rules of the game.

Sheldon has a strong sense of the larger social utility of these typologies. He believes that somatotyping provides a basis for the creation of racial norms (once we have seen the relative distributions of these types among the various races). Following this fanciful use of the concept of race, he states that "we have discovered no case in which there has been a convincing change in the somatotype" (p. 221). Thus not only will we have clear racial types, but they will be unlikely to change over time.

He argues that "by constructing and applying a simple scale for measuring the relative strengths of motivational drives, we have not only found that analogous elemental components appear to express themselves in temperament almost as clearly as in physical constitutions, but we have been able to measure three elemental components with some reliability" (p. 225). Sheldon advocates somatotyping of children because "we need to dispose the influences to which children are exposed in such a manner that youngsters set their hearts upon values which represent the fulfillment of their own constitutional potentialities" (p. 227).

But what does this mean? On the one hand, Sheldon has stated that there is a direct connection between somatotype and temperament and that somatotype does not change. On the other he states that "there is little question that some change in external and manifest motivation takes place in response to educational and environmental influence" (p. 226). He thus categorically states one proposition, then its opposite, and asserts that both are true.

This is not some special defect in Sheldon's thought. This ambiguity has already been noted in the arguments of Hippocrates, Bodin, and Galton. In all these views, the material influences of the genealogical principle and of the direct environmental principle come into conflict and produce contradictions. For Sheldon these contradictions open up a major field of social manipulation. Those who make up the somatotypology and apply the method classify children into categories with fixed behavioral characteristics. They then assert that the children must adjust to their "natural" constitutions, although some modifications can be made. Just what modifications can or cannot be made are up to the somatotyper to determine.

Just as in the related case of IQ measurements, the somatotypes are seen as a system of relatively fixed capacities. Children of the wrong type—endomorphs, for example, should adjust their hopes to lives of uninventive toil. To try to educate children beyond the limits of their types is to lead them into lives of frustration.

Sheldon's methodology, like Kretschmer's, masks a process that brings the humoral typology of human characteristics forward again in new language. There is a direct correspondence between Sheldon's somatotypes and humoral classifications.

His claim to be using the results of modern biomedical science is false. By "biological" Sheldon means only that which is fixed or unalterable. Since key elements in modern evolutionary biology are that biological processes and categories are in continuous flux, that new variation is continuously generated, and that no class of organisms is fixed in any ultimate sense, it is clear that Sheldon's biology is nonevolutionary.

National Character and Ethnic Identities

Attributions of national character are yet another application of the same basic humoral/environmental theories to the explanation of the behavior and historical trajectories of nation-states. The literature is voluminous, diverse, and theoretically flaccid. It ranges from quantified nonsense to chauvinistic political tracts to Sunday travel-page homilies. The theoretical structure of national character arguments has hardly changed in the whole of Western history (Caro Baroja 1970).

I do not propose to delve deeply into this subject here. I only want to show that this major tradition of interpretation of history and justification of conduct rests on a humoral/environmental foundation. Before making this point, I shall evoke the flavor of this literature with examples from Spain, the country where I have done my own anthropological research.

National character attributions regarding the inhabitants of Spain began very early. Strabo was making assertions about the inherent character of the Iberians in the second century before Christ (Caro Baroja 1970). Both Muslim and Christian rulers exhorted their subjects with attributions of national character and destiny during the long battle over the Iberian peninsula. Perhaps most striking of all is the elaboration of the "black" and "white" legends centering on assertions that religious intolerance and greed, on the one hand, and strength in the Christian faith and military virtue, on the other, were intrinsic, natural characteristics of the Spaniards.[3]

The most notable thing about these national character views is the way they support political positions. At times we are told that the Spaniards are fanatical, avaricious, authoritarian, anarchical, or too proud to be governed. At other times the Spaniards are pictured as steadfast in the faith, ascetic, proud, and militant. Negative attributions are generally justifications for political action against Spain or for the failure of some group within the government to fulfill its promises. Positive attributions generally serve to justify governmental policy or the suppression of minorities, or simply to whip up national pride. This is still true in Spain and elsewhere.

Often national character portraits take on literary trappings. In his widely read book *The Spanish Temper*, V. S. Pritchett recounts the following episode:

> Hendaye: the train dies in the customs. One gets a whiff of Spanish impossibility here. A young Spaniard is at the carriage window talking to a friend who is on the platform. The friend is not allowed on the platform; what mightn't he be smuggling? The gendarme tells him to go. The Spaniard notes this and says what he has to say to his friend. It is a simple matter.
>
> "If you go over to see them on Wednesday tell them I have arrived and will come at the end of the week." But if a bossy French gendarme thinks that is how a Spaniard proceeds, he is wrong. The simple idea comes out in this fashion:
>
> "Suppose you see them, tell them I am here, but if not, not; you may not actually see them, but talk to them, on the telephone perhaps, or send a message by someone else and if not on Wednesday, well then Tuesday or Monday, if you have the car you could run over and choose your day and say you saw me, you met me in the station, and I said, if you had some means of sending them a message or you saw them, that I might come over, on Friday, say, or Saturday at the end of the week, say Sunday. Or not. If I come there I come, but if not, we shall see, so that supposing you see them . . ." Two Spaniards can keep up this kind of thing for an hour; one has only to read their newspapers to see that they are wrapped in a cocoon of prolixity. The French gendarme repeats that the Spaniard must leave. The Spaniard on the platform turns his whole body, not merely his head, and looks without rancour at the gendarme. The Spaniard is considering a most difficult notion—the existence of a personality other than his own. He turns back, for he has failed to be aware of anything more than a blur of opposition. It is not resented. Simply, he is incapable of doing more than one thing at a time. Turning to the speaker in the train, he goes over the same idea from his point of view, in the same detail, adding personal provisos and subclauses, until a kind of impenetrable web has been woven round both parties. They are aware of nothing but their individual selves, and the very detail of their talk is a method of defeating any awareness of each other. They are lost in the sound of their own humming, monotonous egos and only a bullet could wake them out of it. [Pritchett (1954) 1965:7–8]

Pritchett has given us, in literary form, a pat explanation for the decline of the Spanish empire and for the Civil War. Spaniards are so egotistical that they are incapable of assessing real-world situations and therefore act with utter irresponsibility. The necessary result is armed violence. Spaniards are to blame for Spanish history.

This genre of writing is familiar to us all. A kind of extended Polish joke, such views are generally associated with serious inattention to power relationships, political economy, and social conflict.

The relationship between these national character portraits and the humoral/environmental theories is close. National characters have to come from somewhere. How did the Spaniards come to be as they are? There are basically only two answers. The Spaniards are as they are either because of the physical characteristics of the Iberian penin- sula or because they were created (by God or some other initiatory principle) that way. Subsequent history subjects them to the influences of other people who come to the peninsula, but the "natural frame" persists nonetheless.

Among the most frequently cited causes of the supposed anarchism of the Spanish character and therefore the political disorders that have occurred in Spanish history are the highly broken topography, the great regional ecological diversity, the difficulty of communication be- tween the political center and the peripheral regions, and the suppos- edly radical distinction between "wet" and "dry" Spain. Geography determines national character and history.

No matter which of the particular geographical features is given primacy, the strategy of explanation remains the same. It is familiar to us from the works of Hippocrates and Bodin. If the physical envi- ronment determines national character, there must be a way that en- vironment gets inside of the inhabitants to have its effect, to become part of them. Thus national character attributions in which geogra- phy plays an important role, whether they be pre- or post-Darwinian, rely on a logic that belongs to the humoral/environmental theory. People are a product of their environment, not in the evolutionary sense of adaptation, but in a humoral sense: the environment directly deter- mines the humoral balance in the population, and this balance in turn determines the national character.

Often this environmental determinist position is found in a confus- ing mix with some kind of genealogical myth. Using the Bible as a source of information, many national character writers have tried to connect the contemporary population of Spain, or of any other European country, with biblical figures. At the initial creation of the divergent groups of human beings, the fundamental differences between them were set. Subsequent history moved them around the landscape until the modern nations emerged. In this view the genealogical principle is clearly central. Nationalities were created and have "bred true" since then. The only changes in their behavior have been wrought by the environments into which they have been forced to move.

These environmental and genealogical views are often combined to create an ambivalent explanation of national character. Eléna de la Souchère approximated Hipprocrates' view when she wrote in 1964 regarding the fatal flaw of excessive individualism in the Spanish character: "This innate individualism was to become more specific, emphasized by the environment and, from the first, by the long struggle of Spanish men against an arid land" (la Souchère 1964:15). This is the very ambiguity that characterizes the views of Hippocrates, Bodin, and the great mass of thinkers who seek to explain social structure and political behavior on the basis of "natural" differences between human beings.

Conclusion

The catalogue of thinkers could go on almost indefinitely. The controversies over immigration quotas and the outbursts of racism in the 1930s and 1940s are well known to us all. The underlying conceptual structures have much in common with those discussed here. Thus I conclude that in the vast recent literature on races, national characters, constitutional types, and eugenics, pre-Darwinian humoral/environmental ideas prevail, despite regular attempts to claim the support of evolutionary theory.

Our cultural tradition finds it exceedingly attractive to anchor explanations of behavior of groups of people in an original place of creation. That place may be in the hand of God or in a particular physical environment, but it stamps the "nature" of a people forever.

To the extent that subsequent modifications occur, they occur only as alterations in that basic "natural frame."

What predominates is a view of different races, nations, constitutional types, and social classes as fixed categories with little internal variability and no capacity to produce future variability. All of these theories reject so-called mongrel alliances because mixture of pure types generally causes social degeneration.

In view of evolutionary biology's emphasis on the continuous production of variability and the complex nondirectional dialogue between the variability produced and the variety of environmental pressures to which these variants are subjected, it is clear that the claims of any of these theories to be evolutionary or or to represent the findings of modern science are false.

III Complex Continuities

The case for simple continuities of pre-evolutionary thinking in the works of such authors as Galton, Kretschmer, and Sheldon is easily made—so easily that many critics of contemporary biological determinism dismiss important contemporary authors with a polemical but perhaps too derisive wave of the hand. The Sociobiology Study Group, for example, dismisses E. O. Wilson's statements about humans in an article titled "Sociobiology—New Biological Determinism" (1977).

Such an approach will not win the day. There are indeed continuities between pre-evolutionary and contemporary biological determinist views of humans, but the issue is not therefore simple. A number of contemporary thinkers have made good-faith efforts to treat humans in accordance with what they conceive to be the requirements of an evolutionary perspective, only to continue key elements of pre- and nonevolutionary thought against their own intentions. This complex kind of continuity can be seen in E. O. Wilson's human sociobiology and Marvin Harris' cultural materialism, among many other contemporary biologically based approaches.

Two mistakes are often made in attacking such efforts. The most common is to assume that these scholars are acting in bad faith, that they are nefarious conspirators against the truth. The reduction of scientific controversy to a battle between the good guys and the bad guys is a strategy usually employed when intellectually based criticism seems too weak for the task. To assume that such thinkers are operating in bad faith is both intellectually sterile and overoptimistic. More frightening to contemplate is the possibility that intelligent, sincere, well-trained scientists can expend their efforts on these problems and then unknowingly reproduce quite predictable pre- and nonevolutionary views. This possibility suggests that a powerful and durable cultural system, rather than some personal failure of rationality and good faith, lies at the center of the problem.

Equally important is the consistent failure to set contemporary thinkers in an adequate intellectual/historical context.[1] Were I now to discuss Wilson and Harris directly against the backdrop of Hippocrates, Jean Bodin, and Francis Galton, I would be treating them stereotypically. Their arguments are detailed, complex, and keyed to a restricted range of biological and evolutionary issues—the evolution of social forms for Wilson and the relation between social practices and ecological adaptation for Harris. To treat them as if they were global biological determinists is to miss the subtlety of their arguments. Such an approach also fails to account for the attractiveness of their arguments to many academic audiences, unless we believe that these audiences also share social class interests with the authors.

To cope with these problems, it is necessary to pose such contemporary thinkers as Wilson and Harris against analyses of some of their legitimate pre-evolutionary counterparts. A fair comparison is one in which the specific subject and the level of detail of the pre-evolutionary view are on a par with those of the contemporary author's effort. For this reason, Chapters 4 and 5 demonstrate in detail how humoral/environmental theories have been deployed socially to explain and legitimate particular social arrangements.[2] In the first case, they are used to support a hierarchical system through a strong emphasis on genealogical arguments; in the second, a more egalitarian system is supported by a view that centers, in part, on assertions about the requirements of successful accommodation to the environment.

In these detailed contexts, which elaborate considerably the arguments I have already made about humoral/environmental theories, I set the works of E. O. Wilson (Chapter 6) and Marvin Harris (Chapter 7).

CHAPTER 4

Purity of Blood
and Social Hierarchy

Humoral/environmental theories have generated highly specific and elaborate rationales and explanations of particular political systems, both hierarchical and egalitarian. Indeed, much of their appeal arises from their promise to correlate detailed and specific "natural laws" with particular social structures.

The ideological system that supported and explained the separation of nobles and commoners throughout the Middle Ages and the Renaissance rested on humoral/environmental theories, with their reliance on fixed "natural" categories. European concepts of nobility were based on the assertion that nobles enjoyed superior social status because of the material quality (purity) of their blood; that is, the social hierarchy expressed a natural hierarchy in the quality of the humors of the populace. These ideas, far from being rigid, could be employed to explain and justify very different kinds of hierarchical social structures. There is a tendency to treat the past in unidimensional fashion. Part of the staying power of humoral/environmental theories arises from their immense flexibility.

Blood has received an extraordinary amount of attention in Western thought. Blood as the primary source of the other humors, blood as life, blood as death, bleeding of patients, menstrual blood, the blood of religious sacrifice, and the blood of kinship encompass a vast field of Western discourse. We can get a glimpse of these riches in the widespread notion of the pure blood of nobility.

Classical Ideas about Blood and Behavior

Beyond the four-part humoral view already discussed, certain classical ideas about body fluids form a backdrop for the concept of blood nobility.[1] In Greek thought, the lungs (viewed as blackish, spongy sacks containing blood and breath) were the seat of consciousness. The various states of consciousness were attributed to degrees and types of moisture in the lungs. Dry lungs yielded the alertness and sobriety characteristic of the waking state. Wet lungs, characteristic of the sleeping state, resulted in loss of awareness and forgetfulness. The drinking of wine could cause the lungs to be wet.

The interaction between blood and breath was the very stuff of consciousness. When air was drawn into the lungs, it interacted there with blood, which gave off its vapors (consciousness and intelligence) in the breath. "Greeks and Romans related consciousness and intelligence to the native juice in the chest, blood (foreign liquids affected consciousness for the most part adversely), and to the vapour exhaled from it, breath" (Onians [1951] 1973:63).

The head—perhaps more accurately the brain and its fluids—was revered as the seat of the seeds of being and individual character. The head was the essence of a person in a genealogical sense. In the Greek view, the head was connected through the spine to the genitals, the two linked by another liquid, the cerebrospinal fluid, called *aiōn*. Together these fluids, blood and cerebrospinal fluid, gave rise to the states of consciousness and essential character of individual human beings. Both *aiōn* and blood wre passed on generationally and both were affected by the environment. Thus both were part of the "natural frame."

The complexity of the distinction between the cerebrospinal fluid and blood gave rise to an extensive medical literature. Learned debates raged about the source of the cerebrospinal fluid and its functions—was it a fifth humor, a product of the blood, or a direct product of digestion?[2]

Regardless of the conceptualization of the relationship between blood and cerebrospinal fluid, there was general agreement that the material states of these fluids directly influenced behavior. The genealogical principles gave an individual a particular constitutional makeup of blood and *aiōn*, and the environmental principle continuously acted on that "natural frame," causing modifications in their states.

In explanations of nobility, the main emphasis was on the primacy of the genealogical principle in the creation of noble behavior. One could be noble by genealogy only. To admit environmental influences on nobility would be to imperil the exclusionary system. Yet humoral theories are by no means intrinsically nonegalitarian, as the discussion of Enlightenment uses of humoral doctrines in Chapter 5 will show.

Concepts of Nobility and Blood in Spain

The wealth of Spanish documentation on the subject of blood nobility is awesome, and the diversity of motivations of the writers adds a fascinating complexity to the subject. Classical authors, churchmen, monarchs and their jurists, and jurists representing other interests were all involved.

The Classes of Nobility

The most widely accepted classifications of types of nobility appearing in Spanish documents from the fourteenth century onward were the product of syntheses developed by Spanish jurists who read the classical and ecclesiastical texts on this subject and then disputed each other in print. According to these authors, there were three classes of nobility. The first, primary natural nobility (*nobleza natural primera*), included all classes of entities, animate and inanimate. Because God created all the categories, they all had intrinsic dignity and importance. Each species of entity contained better and worse representatives. The best representatives were called "noble." The connection between this idea and the chain of being is clear. What is noble in the natural world is that which most closely approximates the eternal Idea of it. This first category of nobility formed a background for all viewpoints and was not actively disputed.

The second class of nobility, natural secondary and moral nobility (*nobleza natural secundaria y moral*), was unique to human beings. It came to individuals either through direct inheritance from the first fathers of humanity or because, through great acts of valor or wisdom, the individuals had restored their bloodlines to the purity characteristic of the first fathers. This class of nobility was also called nobility of blood (*hidalguía de sangre*).

Humans were initially created by God in a state of purity. In this original state, all human actions were right actions, for nothing could have caused them to be otherwise. But humans were also created with the ability to sin, and through sin they fell from this original state of purity. Those humans whose behavior most closely approximated that of the first fathers of humanity and who, through all generations, maintained a steadfast commitment to right actions and reverence to God were considered to be noble: "Nobility is nobility that comes to man by lineage" (Alfonso X [El Sabio] [c. 1265] 1848).

In this view, nobility was the closest approximation to the original purity of creation, and it was transmitted genealogically. Those who through sin, heresy, or disloyalty stained their bloodlines were no longer noble. Such people were, of course, the immense majority.

There were two categories of people who could claim nobility of blood. The first consisted of the *magnates*, those extremely famous and wealthy Spanish families whose background and nobility could not be questioned because of their social power. The behavior and social prominence of another lesser group suggested that they, too, were noble, though they did not have the power and wealth to force public recognition. These people petitioned the ruler for letters patent of nobility (*executorias*). In theory, the ruler could neither absolve people of their sins nor purify their lineages; but as God's lieutenant on earth, he had the power to examine the records of a person's behavior and family background. If these records indicated that the person was truly noble, the ruler could grant the letters patent that "recognized" (not created or granted) that nobility. Nobles who gained their status in this manner were called nobles by letters patent (*hidalgos de executoria*) but were also considered to be nobles by blood (*hidalgos de sangre*).

The third class of nobility was civil political nobility (*nobleza política civil*). This kind of nobility was granted to individuals by a ruler in recognition of their service to the crown. It was a prize of honor awarded by the state to its servants because of their superiority in the use of the sword or the pen. Such people were also called nobles by grant (*hidalgos de privilegio*). There were numerous categories of grant (Isasti [1625, 1850] 1972, Moreno de Vargas [1636] 1795, *Nueva recopilación* . . . [1696] 1918).

Thus there were three major roads to socially recognized nobility: proper genealogy combined with general public recognition of it, proper genealogy and right actions recognized as such by a ruler, and service to a ruler sufficient to merit a grant of nobility. In theory, all three rested on the same basic principle: the genealogical transmission of material purity of blood that caused right action and belief. The purity/nobility relationship was the core of this naturalistic explanation and justification of human behavior and hierarchical social structures.

Double Meanings

A key to the operation of this system of concepts was the multiple meaning of biological/physical terms. Blood was a physical substance circulating through the body and, following the humoral theory, was a direct cause of an individual's character and actions. Certain qualities of blood were important in the concept of nobility: purity, clarity, and cleanliness. It was not blood itself that made right actions, but its purity, clarity, and cleanliness. Purity of blood was not conceived as a metaphor in any sense; it was felt to be a specific physical property. Purity of blood resulted from genealogy and consanguinity.

The antitheses of these concepts helped to bound this conceptual universe and set its social context. The opposite of nobleman was commoner, and the opposite of the nobility was the populace. The quality opposed to purity/clarity/cleanliness was impurity or (the term most commonly used at the time) mixture. The opposite of nobility was thus mixture, meaning both physical mixture of noble and non-noble blood (creating impurity) and the social mixture arising from unknown genealogical background (always assumed to mean mixed noble and commoner elements). By the same logic, the state of purity had to be proved, for purity was the exception. The ordinary human condition was mixture.

A number of ambiguities must be dealt with at this point. First, as we have seen, there are two Spanish terms that we translate as "nobility" in English: *nobleza* and *hidalguía*. My understanding is that *hidalguía* came into use later and that the term emphasizes the social implications of nobility. The derivations of these terms supplied by jurists of the period are highly fanciful.

Ambiguities in the meanings of blood do not end with nobility, since ideas about blood expand into the realm of fertility, racial differences, and so on. There were also complex debates about the nobility of women, especially when a noblewoman married a commoner or a commoner woman married a nobleman.

Principles and Social Realities

A much deeper ambiguity centers on the sources of nobility themselves. In the ideal model, nobility was a direct genealogical transmission from the first fathers of humanity, who were created pure in blood. By this genealogical principle, anyone who was noble had to be directly descended from them. Yet the theoretical systems also recognized the possibility that people could, through right acts, restore purity to their bloodlines. This view is much harder to rationalize theoretically within the genealogical principle. After all, if purity of blood directly caused noble behavior, how was it possible for someone with impure blood to act in such a way as to purify it? The difficulty is great and its logic is readily understood. The legitimacy of noble privilege was given a naturalistic justification in a genealogy that supposedly placed it beyond the reach of most people. After all, a privilege that anyone could receive would be no privilege at all. Thus the whole idea of nobility was tied to the genealogical principle.

Yet a social system that could not accommodate social mobility could not survive. The active and often wealthy servants of the monarchy who were not noble had to be dealt with, even at the expense of logic. The idea of royal "recognition" of nobility was an attempt to paper over the granting of noble status to nonnobles. It covered the breaching of the system by claiming that these new nobles had been noble all along but memory of their genealogy had been accidentally lost. Thus in a society in which the efficacy of the idea of nobility as the legitimation of inequality depended on the genealogical principle, people were becoming noble all the time. And by 1600, noble titles were being bought and sold.

Impurity also was fraught with ambiguity. In one sense, impurity was the expression of human sinfulness, something created in our original "nature." Here it had a genealogical sense. Yet nobles were, theoretically, always in danger of losing their purity. But if purity of blood directly imparted nobility to behavior, how could behavior leading to impurity arise? Social reality had to be dealt with. Any social ideology that does not allow powerful people to fall from preeminence is exceedingly vulnerable. There had to be an idiom for downward social mobility as well.

These ambiguities in argument about purity of blood and nobility reflect, in part, the necessity of adjusting a theoretical system to the complexity of a real society. While the legitimacy of nobility rested fully on the genealogical principle, the system had to accommodate the rise of nonnoble families and the fall of noble ones.

All the theoretical contortions notwithstanding, the genealogical principle could not account for social mobility. When egalitarian doctrines came to prevail in Europe, they did so, in part, by forcing this problem to its limits. If the nonnoble could rise and the noble could fall, then the environmental principle, not the genealogical principle, was the paramount force in society.

The continual tension between the genealogical and environmental principles is a fundamental characteristic of the humoral/environmental system for explaining "human nature." The two principles contest each other's turf but neither can displace the other. In Hippocrates they collaborate; during the Old Regime studied here, genealogy is argued against environment; and during the Enlightenment, as we shall see, environment comes to be argued against genealogy. And this nature/nurture debate has not yet ended.

The Social Context of Nobility

According to the great lawmaker and compiler of legal codes Alfonso X (El Sabio) ([c. 1265] 1848:vol. 1, Title XXI), society was naturally divided into three estates: clergy, military, and laborers. The first two estates were noble by definition; they could not have been otherwise, since it was their nobility that made them preeminent in religious and military matters. To hold a position of significance in the church or in the military, a person had to be noble. Within the nobility, there were distinctions of reputation, wealth, and power.

By the fourteenth century these principles were embodied in characteristic social institutions and patterns of social conflict. The documentation of claims to nobility became an extremely important function of the state; heraldry and genealogical investigation flourished as never before or since. All families that could make claims to nobility did so and insisted on the issuance of letters patent.

Because of their crucial role in military actions and governance in the late part of the Reconquest in southern Spain, the military orders (originally established for the Crusades) came to exercise important control over the process of granting letters patent. Ambitious individuals with sufficient wealth to receive a proper education found that admission into the military orders was a vital step. To secure admission an individual had to prove his nobility.

If the applicant did not have an established claim to nobility, the military orders instituted a complex investigative process. Genealogical research was undertaken to ascertain that there was no Moorish, Jewish, or heretic mixture in the man's background. Testimony was sought from acquaintances regarding his behavior, and investigators visited his town of origin to see his properties and to discuss his reputation with townspeople, especially to see that neither he nor his family had engaged in nonnoble occupations. The investigative panel then determined the nobility of the applicant, subject to royal confirmation.

There were other avenues to the social recognition needed for ascension to higher statuses (preeminently through the church), but this example suffices to show how fully developed the administrative/legal apparatus surrounding grants of nobility was. The gatekeeping function of nobility was considerable and carefully exercised.

In any such institution, great opportunities for abuse exist. Enemies could make false claims about an individual's background and people could falsify their own claims to nobility. By the seventeenth century, letters patent and privileges were easily bought and sold. With enough money a person could become noble by either bribery or direct payment for a title (Caro Baroja 1966).

The nobility gained center stage with the beginning of the definitive administrative centralization of Spain in the fifteenth century. The issuing of letters patent, the development of complex rules for dealing with nobility, and the elaboration of legal concepts of nobility began to appear in great numbers by the time of Henry IV (1454–74). A significant number of claims to nobility were considered by Philip II (1527–98) and his successors. Philip was particularly concerned with the problem of recognizing the "native" nobility in countries then incorporated in the Spanish empire. He used grants of nobility as part of a strategy of political alliances to operate his highly heterogeneous realm.

The central role of nobility effectively came to an end by 1700, when wealth became more important than titles. Once wealth could purchase nobility unproblematically, the social value of nobility began to decline (Caro Baroja 1966). This is not to say that society was becoming less stratified by 1700; rather the idiom of stratification was shifting from an aristocracy of blood to an aristocracy of wealth. Although wealth had obviously always been important to a family that aspired to nobility, by this time wealth alone, unadorned by title, could provide great social eminence. The social power of the genealogical principle had been undermined.

Blood and Nobility in the Basque
Country and Castile

Despite a shared conceptual framework and the use of the same literary and legal sources, the various regions of Spain appropriated and developed the concepts surrounding nobility in different ways. All accepted the causal role of purity of blood in creating existing social arrangements. Despite their use of common concepts for thinking about and legitimating social structures, regional political economy and the course of historical events ultimately differentiated the Basque Country and Castile so completely that they developed mutually opposed political ideologies. This point, central to the understanding of the process of ethnogenesis, shows that, even within an agreed-upon conceptual framework, highly differentiated social implications can be drawn from the same humoral/environmental theories.

Nobility in the Basque Country

The Spanish Basque Country is composed of the provinces of Vizcaya, Guipúzcoa, and Alava, to which Navarra is often added, all in the northeast corner of Spain. There are three other Basque provinces on the French side of the border. Speakers of a language unrelated to any other currently used in the world, the Basques have been in the news in recent years because of the political strife surrounding them, especially the waves of ETA terrorism. The Basque Country has a long history of provincial customary law and has characteristic institutions for local and regional government. The issue of nobility in the Basque Country was first joined in the development of these bodies of customary law.[3]

The conflict occurred in a resolutely political context: the forced development of a comprehensive written code of customary laws (*fueros*) acceptable to the Spanish monarchy. Until the time of this codification, the reigning monarchs had generally observed the customary laws without having a compilation of laws to refer to. The result was continual friction, since the actions of the monarchy were regularly held to be in violation of the *fueros* by the Guipúzcoan government and the monarchy always suspected that the Guipúzcoans were establishing legal precedents to avoid complying with royal wishes.

These laws were written down only when the persistently divergent interests of the crown and the province threatened their very existence. Apparently the first attempt to write down some of the *fueros* in Guipúzcoa was made in 1397. Subsequently the body of written *fueros* was elaborated and recompiled as the changing political situation warranted, until in 1696 the crown demanded a definitive compilation to which no more could be added. These bodies of customary law, unwritten and then written, were of such importance that new monarchs, upon taking the throne, had to swear to uphold them.

The specific history of the *fueros* of Guipúzcoa is less important here than their political context. The compilations were organized as attempts to develop a comprehensive doctrine of provincial rights with legitimating philosophical and legal arguments. The written *fueros* were essentially defensive documents, and they became more complete and strident in their claims for provincial rights as the monarchy's desire to eliminate those rights increased. The maintenance of these *fueros* into the late nineteenth century was permitted by the unique strategic importance of the Basque Country in Spain (Greenwood 1977).

In this defensive process, Basque jurists both compiled laws and set them in a comprehensive historical, geographic, and ethnographic context. A ruler who swore to uphold the *fueros* simultaneously ratified the Guipúzcoan view of history and ethnic identity. And in this view of history, the concept of nobility occupied a central place.

Nobility appeared early in the compilation in the second chapter of Title II. The document contains a comprehensive theological, legal, and historical argument, peppered with references to earlier thought on the subject. The three types of nobility were specified in detail and then an important step was taken. The Guipúzcoan Basques claimed that they were all *hidalgos de sangre* because anyone born in Guipúzcoa of Basque parents was noble:

Among all these types of nobility, that which really and truly refers to the founders of the Province of Guipúzcoa is the natural secondary, which is commonly called nobility of blood, because it is nobility that comes to persons through lineage. This honor comes to them by right and justice via inheritance from the first fathers of humanity. Although there are authors who with some basis assert that all nobilities originated in concessions by kings and natural lords, this general proposition does not fit well with the true origin of Guipúzcoan nobility, which . . . is general and uniform in all descendants of its territories, without having been conceded by any of the kings of Spain, as is manifested by the lack of memory of such, or acquired by the means provided by law, or transplanted here by any of the many foreign nations that dominated this kingdom (since there would have been a historical record of it), but rather is conserved and continued from parents to children, inviolably from the first inhabitants of the Province to the present time. . . . [*Nueva recopilación* . . . (1696) 1918:18][4]

This interesting claim to collective nobility had multiple implications and justifications. The supporting arguments offered were mainly historical. The compilers of the *fueros* claimed that the Basques were the oldest inhabitants of Spain and were lineal descendants of Tubal. The second title of the *fueros* states:

About the beginning of the populating of Spain after the universal flood and about the location in which the descendants of the patriarch Noah first formed their habitation and home one finds no definite information in the gospels: but such [information], which is greatly detailed and strongly based on common authority [popular memory], exists [stating] that Tubal, fifth son of Japheth and grandson of the second father of humanity, was the first who came to this region from Armenia after the confusion of tongues in Babylonia, with his family and others, and that his first settlement and home was in the lands situated between the Ebro River and the Cantabrian Sea. . . . [P. 14]

Thus they argued an unbroken genealogy back to the fathers of humanity, an argument widely made in other tracts as well (Echave [1607] 1971, Isasti [1625, 1850] 1972, Zaldivia [1517] 1944). The physical purity of blood is an important element in this view. After the Flood, humanity had been purified of all but original sin. Those who could claim an unbroken genealogical connection to such figures as Tubal could assert their purity of blood and thus their nobility. Genealogy here attests to humoral purity.

The *fueros* also argued that the Basque area had never been overrun by the Moors, and that the Basques not only defended the area against them but were active participants in the Reconquest of Spain. Claiming to be widely known for their staunch Christianity, they used these combined religious and military arguments to support their claim that their genealogy went back to the beginning of time and that it had never been contaminated by Moorish, Jewish, or heretic mixture.

The *fueros* and the supporting commentaries did not stop with this historical argument for lineal purity of blood. All of the sources carefully documented Basque participation in the Reconquest and encounters with the French (including Roncesvalles) and many others, all showing the preeminence of the Basques in military struggles. Great detail regarding the battles themselves was given, and most royal oaths to uphold the *fueros* mentioned Basque military prowess. The religiosity of the Basques and the large number of learned men the Basque Country had produced were also documented. Thus the *fueros* approached the proof of nobility from the side of right actions as well as genealogy. By their consistent right actions, the Basques legitimated their claims to nobility of blood.

Collective nobility is a most peculiar idea. A major function of nobility is to exclude most of the population from participation in elite institutions. The Basque claim to collective nobility by purity of blood forced the Basque jurists to argue that many people whose social roles would be direct impediments to nobility—farmers, fishermen, coopers—were noble. The Basque jurists pursued this position aggressively:

> It should be noted that nobles of *blood*, particularly those of Guipúzcoa, do not lose their nobility through working in ordinary and necessary occupations, even if they have fallen into total poverty; because nobility of blood did not arise in them but came to them from their ancestors and lineage, and it is enough that it [nobility] should have produced its effect in the former even though at present it has ceased to do so. . . . But if the nobility is nobility of privilege, which is called *ex accidenti*, it is lost in the exercise of ordinary occupations. . . . It should further be noted that the noble who lives nobly, even if he is a rustic and works with his hands, does not lose his nobility. . . . [Isasti (1625, 1850) 1972:47; emphasis his]

This argument represents a fascinating play on the genealogical principle. To place their claim to nobility beyond the historical reach of the Spanish rulers, the Basque jurists argued that Basque nobility was a direct unsullied inheritance from the first fathers of humanity. But in the empirical world Basques necessarily occupied all social strata, many engaging in nonnoble occupations—a direct contradiction of the concept of nobility. To deal with this problem the jurists suspended the behavioral side of nobility entirely and stressed only direct genealogical connections to the first fathers of humanity. This strategy shifted the social function of the concept of nobility from an explanation of social stratification to a legitimation of regional ethnic rights.

The Basque claims were not made or taken lightly. The theorists of monarchy, even as far back as Alfonso X in the thirteenth century, argued that only a very few people, and perhaps no one currently, enjoyed nobility unless it were confirmed or granted by the monarchy:

> There have been and are many who received nobility solely by being from particular territories and places that were noble by privilege and grace from Kings and Princes. . . . [He gives the example of the Roman cities in Spain.] The reason that these and other similar cities and places received this nobility was that they deserved that the Kings and Princes should concede it to them for the virtue, valor, and services lent by their inhabitants. . . . In this way the Vizcaínos [Basques], because of their great antiquity and invincible force and because of their heroic military actions, have acquired nobility for their country, in such a way that by only proving that they are original inhabitants of Vizcaya [the Basque Country], or descendants of such by legitimate and natural male lines, they receive letters patent of nobility of blood, because they truly are [noble] and are declared to be such, this nobility being confirmed by the Kings of Castile and León. . . . [Moreno de Vargas (1636) 1795:30–31]

The monarchy argued that any special privileges the Basques enjoyed must have been given to them by rulers. In other words, the rulers rejected the Basque claim to nobility of blood without royal confirmation, thereby rejecting the Basque claim of a unique ethnic identity. Had the Basques accepted this royal view, it would have been only a short step to a royal argument that Basque nobility was really only civil political nobility and could be revoked by the monarchy. In their counterargument, the Basques claimed to have been noble long before there were any Spanish kings to grant nobility.

At stake in this argument was an important political principle. If Basque nobility were subject to royal confirmation, and particularly if it were defined as civil political nobility, then the ruler who had confirmed or granted it could conceivably choose to revoke it. Arguing that their nobility was natural secondary and moral nobility without need for confirmation, the Basques moved politically against the rights of the Spanish rulers to exercise unconditional political power in the Basque Country. For hundreds of years this argument was a major ideological support to the demand for a semi-autonomous administrative regime in the Basque Country which would operate on the principles embodied in the *fueros*.

This is not to say that social hierarchy was absent in the Basque Country. The Basque Country was as socially stratified as the rest of Spain. Many noble Basque families that shared the collective nobility of all Basques were also civil political nobles with personal privileges that had been granted by Spanish rulers. Social classes and social conflict were certainly not unknown in the Basque Country. To romanticize the Basque past on the basis of a literal reading of the *fueros* is an error, a point amply developed in an unfortunately uneven book by Alfonso Otazu y Llana (1973).

The pro-*fuero* argument was not that all Basques were socioeconomically equal but that all Basques shared equally in a noble genealogy and the rights that arose from it. Collective nobility stressed the genealogical principle in its most radical form and treated the differing social positions of Basques as accidental environmental effects. In the Basque view, the concepts of nobility by virtue of purity of blood and collective genealogical equality were directly linked. Collective nobility became a naturalistic justification for a particular set of political arrangements in which the Basques were singled out for special treatment.

Nowhere is the manipulation of these principles clearer than in the *Corografía . . . de Guipúzcoa* of Father Manuel de Larramendi ([c. 1754] 1969). Writing when nobility as a social ideology was on the wane and egalitarian ideas had begun to spread, Larramendi shifted the ground of the debate to stress the egalitarianism inherent in the idea of collective nobility and a representative form of government under the system set up by the *fueros*. All Basques are equal, he argued, because all are descendants of the same ancestors; and they are superior to the Castilians because they are all genealogically noble and Castilians are not.

The intellectual foundations of the democracy he advocated are not those of contemporary democratic thought. Among Basques a unique degree of human equality was asserted to exist. People from all stations in life had similar claims to human dignity, claims supported by the Basques' reading of the egalitarianism of Christian salvation. But this equality existed, theoretically, only because of purity of blood, because of the unsullied Basque genealogy. Thus the Basques saw themselves collectively as an elite. Their only equals were the monarchs and nobles of the rest of Spain. The common people of Spain were inferior to them because their blood was impure. The genealogical principle here takes a racial turn. Perhaps the most apt comparison is with the 'democracies" of antiquity, which limited participation to a certain group of people.

Basque collective nobility was an extremely difficult problem for the monarchy. In swearing to uphold Basque customary laws—as the Spanish monarchs repeatedly did in an attempt to contain their fiscal and military problems—the monarchy actually ratified the Basque view of history. Royal subjects in other parts of Spain complained bitterly about Basque rights, arguing that such rights should either be extended to all subjects or withdrawn from the Basques. It was not until the latter part of the nineteenth century that the *fueros* were officially canceled. And now, with the new constitution, the *fueros* are once again a political reality.

Nobility in Castile

Castile is the central region of Spain, made up of the provinces of Avila, Burgos, Logroño, Santander, Segovia, Soria, Valladolid, Palencia, Ciudad Real, Cuenca, Guadalajara, Madrid, and Toledo. Its historical trajectory was quite different from that of the Basque Country. Not only did Castile dominate Spain politically from the late fifteenth century on, but earlier it was almost completely overrun by various Muslim groups (called Moors in the literature). Parts of it were under Muslim control for periods of from 150 to 700 years. A substantial Jewish presence in Castile is also well documented (Caro Baroja 1978).

The ink and blood spilled over the Moors, Jews, Old Christians, Moriscos, Christianized Muslims, and converted Jews is familar enough.

It is important to understand, however, that the arrival of the Muslims and the subsequent intermixing of populations forced the question of nobility to take a distinctive form in Castile. Except for a very few preeminent families with wealth, property, and documents sufficient to forestall questioning of their ancestry, virtually no one in Castile could simply assert nobility by virtue of genealogically transmitted purity of blood.

In this historical context, the role of the confirmation of nobility of blood through letters patent and the granting of civil political nobility became tremendously important. By judicious use of these powers, the monarchy could partly control the powerful and militarily dangerous families that were competing with the crown for power. At the same time, the Reconquest offered a field of honor on which wealthy and ambitious men could do battle. Through their valor they could win the gratitude of their rulers—gratitude that took the form of letters patent or grants of civil political nobility.

Thus the Castilian conception of nobility was almost exclusively military. This attitude could be seen as far back as Alfonso X. Of course, such nobility could be won by only a small segment of the population, since wealth, education, and staff were needed to mount a successful military career. Nobility thus became the principal symbol for social hierarchy. While genealogical connection was obviously important and a broken genealogy could eliminate a family from the ranks of nobility, de facto social eminence or military virtue was necessary for a successful claim to nobility.

Once the major noble families were well entrenched in Castile, their various lines quickly came to control both military and religious institutions. They participated in the establishment of bureaucratic procedures for determining nobility and for granting letters patent. These procedures were used effectively as a device to keep nobility and its privileges a significant monopoly of the few against the many.

The privileges of a nobleman were very considerable. The following list is typical:

monopoly of high offices
monopoly of diplomatic positions
monopoly of command at forts and castles

no payment of taxes except for public works of benefit to them
no confiscation of property for payment of debts
no imprisonment for debt
if convicted of a crime, jailed differently from nonnobles
could not be tortured
if called to testify in a legal action, testimony taken at their pleasure
   outside the court
free to refuse challenges to duels from nonnobles
could force the sale of certain properties to themselves
after judges, would receive the best seating at public events
                         [Moreno de Vargas (1636) 1795:Discourse 12]

At least on the ideological level, the confirmation of noble status clearly carried significant social benefits in Castile. How extensively these rights were actually exercised cannot be inferred from this kind of documentation, but the ideal rules show that nobility conferred social preeminence. By implication such documents also indicate the vulnerable social position of the nonnobles in Castilian society.

In the logic of this system, the rulers and their lieutenants occupied a crucial position. Since virtually all nobility required confirmation or was granted as an honor, the control of nobility became a central instrument of monarchical control in Castilian society, helping to forge an alliance between the nobility and the monarchy against the segments of the population that had wealth and power but were not loyal to the crown.

The Castilian system rested firmly on the principle that human inequality was a profoundly important "natural" element in society. Because men were not naturally equal, the rulers, clergy, and nobility governed in the interests of the majority who were their inferiors. In this case the genealogical principle was used to exclude most classes of people from access to positions of power. Purity of blood was here an instrument of social hierarchy, while in the Basque Country it was used as an instrument in defense of regional rights.

Conclusion

The Basque and Castilian views are similar in important ways, despite the major differences in their social application. Not only do they use the same humoral concepts and encounter the same problems created by the conflict between the environmental and genealogical principles in humoral/environmental explanations, but both use purity of blood as a principle of inclusion/exclusion. In the Castilian case, purity of blood excludes all but the few from positions of social dominance. In the Basque case, purity of blood includes all Basques in order to set them apart from and in a position superior to the nonnoble Spaniards. By this kind of logic, though this statement exaggerates the case, the Basques are to most of Castilian society as the Castilian nobles are to Castilian society.

This general picture of social stratification could be duplicated throughout Europe in this period, and references to the relationship between social position and the physical qualities of the blood in the veins of the population can be found in most countries. Without the humoral theory, none of the arguments would make sense.

Thus the humoral/environmental theory both explained and justified existing social systems. These naturalistic ideas were linked to powerful social forces. Basques versus non-Basques, nobles versus clergy versus laborers, nobles versus commoners—all such contrasts were treated as social expressions of natural categories. Each category of people was as it was because of the way it was created and the history it has experienced. Natural nobility did not come into being; it was created and either remained pure or was degraded.

The contradictions in the various views of purity of blood can be understood as expressions of the ambivalence between the genealogical and environmental principles in humoral/environmental theories. Purity of blood automatically caused noble behavior, but then some Basques could be noble but behave like commoners and some commoners in Castile could act in ways that caused them to become noble. These are the contradictions found in the Hippocratic texts. Humoral/environmental theory virtually always involves these contradictory relations between nature and nurture.

There is a very strong emphasis on the notion that unambiguous natural categories of living things (in this case, classes of people) exist. All individuals are, in a sense, simply embodiments of these categories. The categories are static, having been created once and then reproducing themselves thereafter. Most of the conflicts between the genealogical and environmental principles are caused by attempts to reconcile this static view of the categories with the known changes in social status that families and individuals undergo. Finally the "natural" hierarchy, by virtue of its being "natural," is therefore asserted to be morally correct. The step from "is" to "ought" is made without trepidation.

Concepts of blood and purity in no way exhaust the uses of humoral/environmental ideas to explain and legitimate social systems and human behavior. Bile, both black and yellow, and phlegm also have long and interesting histories. But the working out of the concept of blood suffices to indicate the structure and importance of humoral/environmental theories as naturalistic explanations and legitimations of social systems.

CHAPTER 5

An Enlightenment Humoralist: Don Diego de Torres Villarroel

While the specifics of the naturalistic arguments in support of hierarchical social structures are no longer familiar, this use of "natural" differences is. In the twentieth-century struggles over racism and ethnocide, oppressive exclusionary social theories based on "natural" differences between humans have been widely employed. This contemporary social experience has led some thinkers to assert that any theory of society containing naturalistic arguments must be inherently oppressive (e.g., Ann Arbor Editorial Collective of Science for the People 1977).

While it is true that many oppressive social theories contain naturalistic elements, the connection between naturalistic arguments and oppression is not direct. Naturalistic theories in general and humoral/environmental theories in particular are not inherently biased in favor of social hierarchy and oppression. Such views can be and have been used to buttress democratic and populist doctrines as well. Many democratic theories assert that all humans are born equal and *therefore* have equal rights. Is this position any less naturalistic than the racist counterposition that whites are born superior to other races and *therefore* have more rights?

Apologists for the failure of democratic ideas to capture the minds of a great many people often lament that racist/oppressive doctrines are intrinsically more attractive to most people. But if naturalistic arguments were once capable of fueling popular support for social reform, why do they not do so now?

These issues are joined in the ideas of a Spanish Enlightenment thinker, Don Diego de Torres Villarroel, who participated in an important and well-documented struggle over the social and moral consequences of humoral/environmental theories in medical science. Torres mobilized humoral/environmental theories to support popular medicine against the existing state monopoly of medicine and used these ideas to denounce a host of social inequities, thereby demonstrating that these theories are not inherently biased in favor of social inequality.

Humoral/Environmental Ideas in Medicine

Each of hundreds of schools of medical thinkers contributed its own version of the medical meanings of humoral/environmental ideas. This multiplicity of uses is a sure indication of the fruitfulness and importance of these theories.

Galen's development the humoral/environmental tradition embodied in the Hippocratic corpus, with the inclusion of Aristotelian and Stoic elements, gave rise to a complex medical system complete with an elaborate pharmacology. During the Middle Ages the humoral/environmental tradition persisted through the Arabic renderings of the original ideas. The so-called Arabized Galenism of Avicenna is a case in point.

With the Renaissance return to the original Hippocratic and Galenic texts, the stage was set for a long polemic. The filtering of humoral/environmental ideas into Western folk medicine, the variety of medical traditions derived from Arabized Galenism, and the ferment created by the reading of the original texts set the medical world in motion. The competing schools based their ideas essentially on different readings of the same basic texts. When Arabized Galenism is compared with the Galenism inspired by a direct reading of Galen's texts, it appears to be a completely different, radically opposed school of medical thought and practice. Hippocratics, Galenists of all kinds, "physicians," "chemists," and herbalists flourished and did battle. Yet despite this ferment, the pace of change in the medical schools was slow. Most medical schools persisted in complex, philosophically abstract medical theorizing long after real alternatives had been suggested.

These developments set the stage for the Enlightenment attack on institutionalized medical thought and practice. Enlightenment thinkers familiar with the classical texts were persuaded that medical knowledge and practice had to be concrete, empirical, and individualized. The purpose of medicine was to understand an illness and to collaborate with the inherently harmonious forces of nature in getting the patient well. For these thinkers, the abstract philosophical training that doctors received and the regular medical use of strong drugs and bleedings were major causes of death.

The tendency of historians and philosophers to overlook the immense social and intellectual influence of physicians was noted earlier in regard to Hippocrates. Physicians, certainly by the time of the Enlightenment, were viewed as scientists and professionals, and their social and political views carried great weight. Long before Darwin, scientists were considered important social thinkers.[1]

The Life and Works of Torres Villarroel

Diego de Torres Villarroel, born in Salamanca in 1694, was professor of mathematics at the University of Salamanca. But he was more than a professor: he was a qualified medical doctor, a priest, natural scientist, poet, dramatist, essayist, the most famous Spanish astrologer of his time, dance instructor for a brief period, administrator of some of the Duke of Alba's property, and philanthropist.

Torres was an unabashed admirer of his famous literary predecessor Francisco de Quevedo; many of his literary works are attempts to mimic the works of the great seventeenth-century Spaniard. Yet with a few erudite exceptions, Torres' works are largely forgotten. His overpowering satirical style and continual sermonizing make his fifteen-volume complete works less than easy to read. Only two of his works have remained in print since they were published. Since 1977, two more have been reprinted.[2] This fate seems partly deserved, but literary criteria of evaluation have seriously obscured the larger historical lessons to be learned from Torres' more scientific work.[3]

This immense and heterogeneous corpus is a rich lode for the anthropologist and historian of science. It articulates a coherent vision of the world, society, and the human condition in which cosmology, theology, ethics, natural science, medicine, and social criticism are linked. Of particular interest is a strong current of social criticism based on a kind of moral egalitarianism. And in all these dimensions humoral/environmental ideas play a critical role.[4]

Torres' father was a bookbinder. Though the elder Torres was well respected in Salamanca and ultimately served the city in important administrative posts, these are most modest beginnings for someone who was to become a university professor (as Torres never tired of reminding his readers). In 1715 Torres took the first formal step toward becoming a priest, but he did not seek ordination until thirty years later. Torres' literary career began in 1718 with the publication of his first astrological almanac. He published one a year until 1753, earning much fame and money, a fact he liked to point out from time to time.

The year 1718 marked his first visit to Madrid, where he was eventually to reside for some years. There he engaged in medical studies and made powerful friends among the nobles. He completed his medical studies but determined never to practice medicine, a promise he broke only a few times, under the pressure of poverty. In 1726 Torres acquitted himself well in the competitive public examinations to fill the chair of mathematics at the University of Salamanca, but he had to wait eight years until the death of the previous incumbent freed the funds to permit him to fill the post.

Torres was involved in more than his share of conflicts. By his own admission, he was a rascal. In 1732 he was exiled to Portugal for his participation in an event that has not been clarified in his autobiography or by any of his biographers. But he returned in time to fill the chair of mathematics in 1734. Though he evidently was a good teacher, he was a trying colleague. Always conscious of his social origins, he saw himself as an intrusive foreign body in the university. He never tired of attacking the pomposity of the other professors. The university records show that Torres was often absent in Madrid, a point his enemies endeavored to use against him. But the records also show that Torres was regularly entrusted with complex, important duties by his colleagues, duties that required the mobilization of his contacts in Madrid.

In 1745 he finally determined to be ordained as a priest. This decision was accompanied by a severe illness, a point that has given psychologically oriented biographers much to speculate about.

He retired from his professorship in 1751 but remained active in university affairs practically till his death, in 1770. During his retirement Torres saw to the publication of his complete works, an effort amply supported by an impressive list of subscribers. He also became an active supporter of charitable institutions in Salamanca and a member of the household of the Duke of Alba.

Torres' corpus can be divided into three clear categories: cosmology-astronomy-astrology, natural science, and medical/moral works. A large fourth category consists of miscellany. The cosmological-astronomical-astrological works are diverse. From the physical structure of the universe to almanacs and star charts, Torres covered a wide array of subjects. The almanacs contain attempts at weather prediction with frivolous (according to Torres himself) predictions of events. But he sincerely believed in the influence of astral bodies on the earth and on the physical condition of living things, and constructed astral tables for public use.

His natural science interests were also diverse, covering such topics as beekeeping, geology, the causes of earthquakes, the sources and uses of mineral waters, and why a rooster transported across European time zones crows at a particular hour. These works convey great enthusiasm for empirical science and a desire to communicate the findings of natural science to the general public.

Torres' medical/moral works combine medical manuals written for the lay public, literary works with a powerful emphasis on health, and extraordinarily harsh criticisms of medical education and practice.

The miscellaneous works form a bewildering array. Among them is

Torres' well-known autobiography, published between 1743 and 1758. In it he justified himself, poked fun at his own foibles, and evened many old scores. He also wrote lives of saints, plays, poems, and even an essay on bullfighting.

Cosmology and Natural Science

For Torres, natural science's empirical, inductive methods served to disprove many ancient commonplaces and measurably to improve the general quality of life. This part of Torres' corpus is enormous. These works rest on a consistent cosmology in which the structure of the cosmos, astronomy/astrology, and natural science are tightly interwoven. For Torres a theory of the material structure and operation of the cosmos was the necessary context for astronomy/astrology and by extension for the study of natural phenomena on earth, such as earthquakes, the humors, and the habits of bees and roosters. Thus to be understood, his humoral/environmental theories must be placed within this larger context.

In *Cartilla rústica* (Rustic note) Torres portrayed himself as the teacher of a peasant who was to use his new knowledge to improve the quality of life in his home village. To begin the instruction, Torres diagrammed the cosmos as a set of twelve concentric spheres with the earth in the center.

> I showed it to him, explaining with circles the order of the spheres. My good peasant looked at them for a long while, and then he said to me: "So we are inside of Heaven?" "Yes, friend," I answered, "and inside of the air and fire, and everywhere we are surrounded by and united with these elements, each to the other, and then to the heavens; and just as the layers of an onion hold together, so this marvelous machine is maintained by natural virtue." [Torres Villarroel 1794–99, 6:166][5]

Through this conventional image, Torres explained that the cosmos is orderly, hierarchical, and totally permeated by the four elements. The entire system is perfectly balanced.

This general vision was associated also with the complementary view of the macrocosm and the microcosm. In his *Anatomía de todo lo visible e invisible* (Anatomy of all things visible and invisible) Torres took some companions on a fantastic voyage to the center of the earth.

> And by the grace of God we have seen the organic body of the Earth and we have anatomized its principal cavities, which without doubt have a great similarity to the human body; for the surface of the Earth is like the skin or hide covering these cavities or regions: the lapidary or mineral material is the skeleton that supports the musculature or fleshy part of the Earth, like the bones of the human body; the four humors that swim inside of humankind are found here; for what else is salt water but phlegm? What is sulfur but choler? What are these black and toasted potions other than melancholy? And finally, what are the veins but conduits filled with the most precious liquor that arises from the distillations that occur in these depths, just as the stomach digests food? . . . [1:56]

In this way Torres asserted that from the macrocosm to the microcosm, the material basis of all things and the operating principles are always the same.

El hermitaño y Torres (The hermit and Torres) elaborates this notion. The entire system is a harmonious whole in which the four elements play a principal role.

> . . . all things of the world, great and small, be they natural or artificial, must sustain in themselves the four humors. Then each, in larger or lesser degree, must emit the selfsame virtue, and when introduced into our bodies, they will nourish them, purge them, they will cause drowsiness or wakefulness, and they will stimulate all other good or bad, healthy or sickly operations that we all sense—the happy and the afflicted, the young and the old the living and the dead—in our human bodies. For all creation concurs to give us health, illness, sadness, pleasure, life, and death. [Torres Villarroel 1977:180]

This idea was further developed in *Cartilla astrológica y médica* (Astrological and medical note) (1794–99, vol. 6). After setting the cosmological context for humans, Torres presented the "four natural virtues": generative, vital, animal, and natural. The generative virtue is under the influence of Venus, the vital under the control of the sun. Animal virtue is not associated with a particular astral body here. Rather it is divided into two realms, each associated with particular primary qualities of matter. The two realms are the cognitive and the sensual.

The cognitive realm is divided into imagination (hot and moist), fantasy (cold and moist), knowledge (hot and dry), and memory (cold and dry). In the realm of the senses, sight is associated with cold and moist, hearing with cold and dry, taste with hot and moist, smell with hot and dry, and touch with a mixture of all four qualities.

Natural virtue, now as one of the four natural virtues, is divided into the four humors, each under a particular astral influence: blood (Jupiter), phlegm (the moon), choler (Mars), and melancholy (Saturn). The possibilities for metaphorical combination and opposition in such a system are immense. At the same time the macrocosm/microcosm link is pressed to its limit.

Torres' system contains nothing unique or new. He was a firm believer in humoral/environmental theories, and they formed the basis of his cosmology, natural science, medicine, and even theology (in part). These beliefs explain his interest in astronomy/astrology (Torres did not differentiate the two clearly). Within the structure of his cosmology and the universal operation of humoral/environmental principles, astral influences are a logical necessity. If the universe is a set of twelve circles with the earth at the center and if the material principles of all processes are the same, then logically movement in any one of the twelve spheres will influence the others, and the larger (outer) spheres will more strongly influence the smaller (inner) ones. Thus Torres believed in astral influences as a matter of scientific faith.

> All lower bodies depend on higher ones, the earthly on the celestial, and among them they sustain a mutual kinship and obedience. The superior bodies send down a particular hidden active virtue to the inferior ones, and because of this, the humors and elements of the organic bodies of man and beast shift, are altered, become corrupted, or increase according to the position and quality of the stars: and we know this from daily experience, the best teacher of all things. [Torres Villarroel 1794–99, 6:13]

The Human Condition: Theological/Moral Populism

Humans, as both material and spiritual beings, must live at once in two realms that are difficult to reconcile. In Torres' view, the spiritual realm is the more inclusive; the body is a momentary part of the soul. The body, as part of the soul, necessarily must be respected, and this respect must take the form of treatment in accord with the general material principles of the universe. The punishment for failure to respect the body is physical torment and death; for failure to respect the soul, moral anguish and eternal damnation. Failure to respect the body is an important step toward eternal death.[6] Yet even the most judicious attention to the body cannot protect it from death. "My whole body is a portable infirmary of humors. *I am sick.* And ruined by nature. I am sick. That is why I have laughed at medicine for being so foolish as to presume to give health to mortal man" (Torres Villarroel 1794–99, 3:34).

Individuals have different humoral constitutions, react differently to material forces, and have different strengths and weaknesses. Care of the body must always be empirically adjusted to the constitution of the individual.

Torres insists throughout his works that illness is no respecter of social class. He treats illness as a portentous reminder of the fleetingness of moral life and its honors. Illness highlights the need to attend to the soul's business. Indeed, Torres often seems to feel that wealth is great danger, because the rich face more temptations and can afford more vices. Occasionally he romanticizes the simple life of the countryman, who passes his life in hard work and simple pleasures.

Torres' social criticism must not be overestimated. His egalitarianism derived necessarily from his belief that all humans are equal before God, not from a desire to promote social revolution in this world. Still his egalitarianism and social criticism, combined with the repeated references to his own humble origins, at least place him in the intellectual tradition that flowered in the great democratic revolutions.

The work that best stands as a summary of Torres' view of the human condition is *Vida natural y católica* (Natural and Catholic life.)[7] Consonant with his ideas, the book was written as a self-help manual for the general public. The first of its two major sections deals with "natural"—that is, physical—health. Torres describes in great detail hygienic and dietary practices designed to maintain bodily harmony. Here humoral/environmental theory is the key element as he passes from diet, exercise, sleep, and excretion to mental health. General precepts are given, but always with the caveat that they must be adjusted to the constitution of each individual. In the second part of the work Torres takes up the precepts to be followed to maintain spiritual health.

Perhaps the strangest and most interesting work of all is *Los desahuciados del mundo y de la gloria* (Those evicted from the world and from glory). Torres is taken by a devil on three journeys to witness the agony, physical death, and damnation of a variety of people. Both sexes, various social classes, and various diseases are represented with astonishing clarity. In each case the clinical side of physical illness and death is presented in excruciating medical detail. Indeed, this apparently disproportionate interest in physical illness clearly spoils the work for many literary audiences. But this detail is integral to Torres' thought about humanity as a "portable infirmary." The multitude of ways in which illness can attack and the helplessness of medicine must serve to persuade the audience that the only final salvation is spiritual.

Following the clinical portrait of each illness is an equally clinical portrait of the causes of the individual's spiritual damnation. All of the patients are damned, and the portraits of the demonic hosts mimic those of Dante. It appears that the physical neglect that led to illness was a symptom of a deeper spiritual neglect. Thus the message of the two parts of *Vida natural y católica* is repeated. Torres was pitiless and repetitious in his condemnation of degradation of the body and of the soul.

The remedies for these ills are within the reach of all people, rich and poor alike. The rich are particularly blamed for their behavior because they have the resources to live correctly and often do not. The poor can be excused in part for their ignorance, as few writers have directed their attention to humble audiences. Torres' desire to communicate these lessons to the humble is an indication of his moral populism. This attitude becomes most pronounced in his criticisms of medicine.

Critique of Medicine

Torres' attitudes toward medicine were rooted in his humoral/environmental ideas, his profound belief in empirical science, and his populism. No specific element of his critique is uniquely his own. His particular criticisms of medicine and his theoretical points of departure were widely shared, as the writing of Martin Martínez (1748) and Benito Jerónimo de Feijóo (1724–39) attest. What makes Torres' views interesting is his combination of widely shared criticisms of medical education and practice with a consistent attempt to create a "naturalistic" medicine "for the people." Within the total corpus we find an immense array of scathing denunciations of medicine.

> The medicine that is studied in the universities is a vocabulary of terms that sound good and do ill, are worth little and cost much, and they sell us their knowledge so dearly that they generally cost us our lives. [Torres Villarroel 1794–99, 4:200]
>
> I read Hippocrates, Galen, Willis, Sydenham, and the bravest of the old and new schools, and I did not find in them a medicine powerful enough to stop the running of a catarrh. In their books and among the doctors, one finds prescriptions to sell, not medicines for curing. Since I began to realize the little science man has in regard to man, animals, and the mineral and vegetable realms, I lost faith in the Aphorisms and I have decided to die by my diet, which is a doctor and medicine that is both cheaper and less disgusting. [4:199]
>
> "But tell me, is it not possible that they [the doctors] have a certain basis on which to found their conjectures?" "Not at all," I said. "If they could prove their ability to cure even the least serious illness, the *doblones* would not fit in their purses. It is a misfortune and an unhappiness how short is their science, considering how long they have studied the art. And so, when ill, I do not order the most famous doctor to be called, but rather the first that passes by the door; all doctors are good and medicine is bad." [2:345]

A major emphasis in Torres' critique was the weak empirical foundation of medicine. He felt that medical education emphasized philosophical abstraction at the expense of empirical research. The scientific pretensions of the medicine of his time thus were to him intolerable pompositics and genuine physical dangers. He counseled good diet and living habits as the best, and certainly the safest, medicine.

Torres portrayed doctors as a dangerous luxury that only wealthy societies could afford.

> No one knows medicine; it is said to exist, but no one knows where it lives. The doctor is a political fraud who serves to decorate republics,

not to cure illnesses; he attends to the ill but does not cure them; he is a witness to the triumphs of nature, the miracles, and the deaths. So if he is infallible and you, sir, are abandoned on all sides, conform yourself to necessity, finish your trip to the other world, die like a Catholic, not like a savage. [4:197]

The doctor does not cure, he merely witnesses the course of the disease and charges for his observation. And doctors have clients because of the cowardice of people faced with pain and death.

Torres' criticism goes further. Doctors are even active agents of illness.

If you are healthy, to seek the doctor is to solicit all illnesses: if you are ill, it is to seek the greater unbalancing of your humors, and to achieve a dubious relief, you will have to endure evident risks and very notable changes. Believe me that the ills of the body are felt and known to all, but no one can cure them. He who places his confidence in the aphorisms of nature and in temperance will be better cured than he who places his pulse in the hands of doctors. . . . In the hamlets they do not use doctors, and the locals live more robustly and longer. . . . Thus if you call him [the doctor], you well can throw your fate to the winds, prepare your patience, and deliver your stomach to concoctions, garbage, and brews, your feet and arms to the barber, and your body to the parish church. . . . More die attended by a doctor than without medical assistance. . . . You need a confessor more . . . he has the true and undeniable medicine, while for the illnesses of the body, there is no known antidote. [3:391–94]

Torres firmly believed in medical self-help and in the obligation to care for oneself, physically and spiritually. No one can know us as well as we know ourselves; the expertise of doctors is a fraud.

What is important for us to know is clear to all: it is the very science of souls, and in that science only he who seeks his own counsel is erudite. The study of medicine begins with knowledge of our architecture and economy: my body is closer to my own scrutiny than that of another. . . . With no more effort than the prudent appreciation of the voices and shouts of natural reason we will know our ailments and their cures better than the doctor; and we are able to care for ourselves better than he can. [4:7–9]

Doctors are political enemies, permitted by republics for the sake of variety, not out of need. Illness remains in the body, and the doctor comes and goes, and the illness remains until it wastes the humor away or nature, embarrassed and bored by the gravity of the treatments, heals itself. [4:84–85]

Torres' critique of institutional medicine was partly designed to convince ordinary people of their ability and duty to fend for themselves. Not only was institutional medicine bad, but what little knowledge there was did not find its way to the people; it was hoarded as a lucrative medical monopoly. *Vida natural y católica*, *Recetas de Torres* (Torres' prescriptions), *Médico para el bolsillo* (Pocket medical handbook), and the various *Cartillas* (Notes) were all written as medical guides for the general public. This popularizing intention is proudly stated in each work. All are written in Spanish rather than Latin, and the language is reasonably simple and direct. Torres believed that people could be their own scientists, their own doctors, because the relevant knowledge was directly available through empirical observation and simple induction. Empiricism and populism were thus linked.

Torres believed that popular medicine would necessarily have certain characteristics, derived from humoral/enviromental theory. Such medicine would be based on nonradical, nonintrusive treatments that supported the "natural harmony" of the body. Diet, environmental change, exercise, rest, and meditation were the keys. Many of Torres' ideas could pass muster among contemporary holistic and naturalistic practioners.

Occasionally Torres, like many contemporary believers in naturalistic medicine, flirted with the idea that most serious medical problems were caused by the "unnatural" way civilized people live.

Those who dwell in this village are generally of more than medium stature, refined appearance, good color, well built, strong, and happily healthy: this is because those who limit their lives to a simple diet, accompanied by the sweet fatigues of their labors, live eighty and ninety years without the cares of ordinary illnesses and without the damages often incurred in social gatherings, libations, and the liberties of cultured civil society. [5:368]

Although Torres approached popular medicine from many angles, his works on the uses of mineral waters provide the best overall synthesis of his medical ideas. Two monographs on three mineral springs in the province of Salamanca link his cosmology, geology, natural science, and social criticism into a general humoral/environmental vision of the human condition. The first monograph, published in 1744, is titled *Usos y provechos de las aguas de Tamames y Baños de Ledesma* (Uses and benefits of the waters of Tamames and baths of Ledesma). Though Tamames has been abandoned, Ledesma is now the home of one of the largest and most modern spas in Spain. Torres dedicates the monograph to the owner of the lands where the springs are located. This dedication evokes most of his humoral/environmental vision.

> The famous spring . . . is a fertile treasure and an endless mineral source that God chose to place in the territories where your excellency is the legitimate Lady, in order to add good fortune, blessing, and happinesses to your most illustrious house. Its waters are a delicious and most pure balsam, through which those suffering the misfortunes of illness recover the natural balance of their humors, the restoration of their lives, and a robust resistance against the ills, corruptions, and upsets to which our miserable weakness condemns us. [4:230–31]

After detailing the properties of the waters, Torres asserts the reasons for their curative powers: "Water, in my understanding, is nothing less than a liquid powerfully suffused with the virtues of the stars, airs, metals, branches, seeds, animals, and all things visible and imaginable in the lower and higher realms of the world" (4:237).

Each person must be treated individually because people's constitutions differ, and the treatments must be explained in language intelligible to the ordinary patient.

> I do not stop to define, divide, or discourse like the hidebound Physician; nor do I increase the number of aphorisms, examples, or authorities because to do so is to spend time and paper uselessly. As a practical, mechanical, and rigorous observer, I prescribe to the ill, some who must drink and others who must bathe in the waters, a tailored and useful regime, a sure and inoffensive diet, a moderate daily plan during the cure; and for afterward I give them consolations and rational hopes

that help them achieve spiritual health and serenity, calm their apprehensions, and leave no room for melancholy. I also put these precepts in ordinary language and in the clearest doctrine so that even the most uneducated patient can understand and govern his body and its ills with no more doctor or aphorism than those found in the directions on these pages; I have founded the whole utility of this doctrine on this intention. [4:233–34]

. . . ordinary water serves and cures all kinds of people, the ill and the healthy, be they cholerics or melancholics, phlegmatics or sanguinaries, because it was created for all and for all it is prepared, disposed, congenial, and suited to their ills and good health. [4:239]

. . . I wanted to give it to them in writing so that all patients could carry with them a cheap doctor; because not all who go to drink or bathe can bring a salaried doctor with them. [4:234]

He strongly criticized the doctors of the University of Salamanca for not having made the public aware of these waters and their uses.

In 1753 Torres wrote again on mineral waters, in *Noticias de las virtudes medicinales en la Fuente del Caño de la Villa de Babilafuente* (News of the medicinal virtues in the Fountain of the Spring of the Village of Babilafuente). This spring is still in use. Here the same general themes are repeated, and the criticism of organized medicine is even sharper. In a prologue addressed sarcastically to the "Deceased Doctors of the Medical Schools of Spain" he denounces the stupidity and even criminality of systematic medical ignorance of the uses of mineral waters, especially in view of a standing request by the Royal Practical-Medical Council for such information.

To the Members of the Royal Practical-Medical Council of Our Lady of Hope in Madrid. It is also a letter that aspires to be a Prologue. To the Deceased Doctors of the Medical Schools of Spain: Dear Sirs:
Some because they lived lost in the foolish delights of their useless speculations, others because they blindly delivered all their gullibility to the potions, mixtures, and juleps that they found in the prescriptions in their books, and the majority of them because their imagination was occupied with other interests, more important than these trifles of public health, none ever remembered to investigate the virtues and effectiveness of the infinite medicinal springs that the industry and effort of nature created in their territories for the alleviation of many ills. Those

living today, some because they inherited their complexions and certainties with the portfolios and maxims [of their predecessors], others because they presume that study, maturity, and experience are superfluous to their practice because they already have the repertoire of gestures, refrains, and ponderations needed to send the layman to the other shore, nurtured themselves on nonsense and ignorance, believing them to be prodigious truths, and have refused to involve themselves in the examination of these precious novelties, nor have they responded to your letter in which you request information about the origin and constitution of the healthful waters whose currents emanate in their regions. The utility of knowing and using the waters is visible, demonstrable, and advantageous to the world; because in truth, these springs are small, clean, easy, safe, and cheap Apothecaries, in whose fountains and currents are found a marvelous mixture of substances, chosen by the prolixity of nature and free of the impure mixtures and adulterations that are found in the compositions created by the Chemist's whim. . . .

The public (Dear Sirs) is the first and most naked community in the world: it is the pauper, the uninstructed, the patient, and the invalid that is most visible and deserving of our contributions, goods, documents, and efforts. [5:363–65]

These virtually unknown essays on mineral waters clearly reveal the structure of the thought of this Enlightenment humoralist. The material cosmos created by god; the geology of the earth giving rise to airs, waters, and places; the humoral conditions of human life; and the battle against socially oppressive and morally inexcusable manipulations of knowledge are linked in a single, consistent pattern of thought.

Conclusion

This excursion into the works of Diego de Torres Villarroel provides a broad sense of the symmetry and interpretive scope of humoral/environmental theories. Torres' entire system is characterized by consistency and balance. Still Torres' system of thought is clearly nonevolutionary, in the same way as is the system used to explain and justify the preeminence of the nobility. In his view the natural world was created once and for all by God and has not changed significantly since then. His faith that empirical investigation would yield useful results is based on a belief that the Creator is beneficent and that his creation is formed of clearly defined, stable classes of things. The natural order is a moral order by definition.

Torres' thought is also characterized by tension between genealogical and environmental principles. He clearly believes that a great deal can be accomplished by manipulation of the environment, and in that sense he differs greatly from earlier Spanish apologists for the social order. His approach to popular medicine is informed by a belief that sensible diet and lifestyle can greatly improve health. That is, he believes environmental manipulation can have important effects on health. His critique of medicine also implies that people who permit themselves to become overintellectualized (as the academic physicians had done) can lose touch with the principles at work in the world. Poor training could only have the power to create poor doctors if the environmental principle were a potent force.

Together two very different deployments of humoral/environmental theories show that these views have a very broad scope. Humoral/environmental theories are complex, flexible, and diverse—and they also have a pleasing overall integration. At the extremes, they have given rise to arid abstract scholasticism and mindless empiricism. They still hold astonishing power in the Western world, as the currency of some of the concepts used to justify nobility in current racist ideologies and the apparent similarity of Torres Villarroel's medical views to those of contemporary holistic medical reformers both suggest.

One of the most powerful characteristics of humoral/environmental ideas, seen in all the literatures discussed, is that they are constructed to make moral and political decisions seem empirical. These views consistently argue that the social structure or human behavior must follow a certain pattern because "nature" or "human nature" requires it to do so. The connection between analysis of the "natural" world and political and ethical conduct is made to appear direct and scientific.

In order to make this argument appear plausible, it is necessary that the natural world be both static and coherent. The humoral/environmental world is one of fixed categories, of constitutions that tend toward harmony. The categories were created once and for all in the beginning and they cannot change. The "marvelous machine" runs on forever. Indeed, the problem of change in categories appears in the persistent conflict between the genealogical and environmental principles in these theories.

When Darwin succeeded in synthesizing the actions of the environmental and genealogical principles and showed that the continuing origin of species (natural categories) was an inescapable theoretical and empirical conclusion, he demolished the very foundation of humoral/environmental theories. With them also was demolished the apparently easy and obvious connection between natural categories and moral rules. The blow was so sharp and so surgically delivered that many of its implications have yet to be assimilated.

To a surprising extent, the static vision of the world on which the humoral/environmental scheme depended marches on, though it is maintained only at the cost of serious contradictions. Nor is it carried forward only by some fringe group of antiscientific thinkers. Many pre- and nonevolutionary ideas persist in the theoretical and empirical works of major contemporary scholars who consider themselves to be in the forefront of applications of evolutionary principles to the analysis of human behavior. The extent to which nonevolutionary elements invade the work of these scholars will, I hope, show clearly how much remains to be done before the Darwinian revolution can be considered complete.

CHAPTER 6

Human Sociobiology

Given the scope of the polemic unleashed by human sociobiology in recent years, no one can enter this arena without some trepidation. It is a minefield because of the complexity of the biological questions involved and because application of powerful biological models to the study of human behavior simultaneously creates theoretical, political, and ethical difficulties. Precisely because this particular subject attracts so much attention, it is reasonable to believe that it touches directly on fundamental ways in which we conceptualize the relationship between nature and culture.

E. O. Wilson's *Sociobiology: The New Synthesis* (1975), a quasi-textbook modeled in important ways on Darwin's *Origin of Species*, gave the field its public identity and set the boundaries of the current debate. That its sophistication in certain areas has been quickly surpassed is not an argument against its general significance. Such a book does not have a great impact simply because of the force of the ideas presented; they must be presented in an order and context that are themselves compelling. *Sociobiology* is a compelling work in this sense.

In *On Human Nature* (1978) Wilson attempts to address his critics and to expand the arguments advanced in *Sociobiology*. Wilson and Charles Lumsden's *Genes, Mind, and Culture* (Lumsden and Wilson 1981) attempts to specify the theoretical framework supposed to be implicit in *On Human Nature*. Though it makes certain points from the previous books clearer, it does not fundamentally alter the structure of Wilson's discourse on the relationship between nature and culture.

There is no doubt that sociobiology has an important contribution to make to evolutionary biology as a whole, no matter what the verdict about human sociobiology is. Ever since Darwin's *Origin of Species* there has been a recognized need for an evolutionary analysis of social behavior. Observations across wide ranges of species show that certain forms of self-sacrificing behavior are common in the animal world and are often advantageous to the fitness of the collectivity though they necessarily reduce the fitness of the sacrificing individuals. By evolutionary logic, such individuals would be less and less represented in populations over time, and this kind of collectively useful behavior would disappear.

Darwin himself was aware of this problem, as his statements on neuter and sterile groups within a species demonstrate. Solutions to it were not forthcoming. It resurfaced with considerable impact when V. C. Wynne-Edwards published his *Animal Dispersion in Relation to Social Behaviour* in 1962. He claimed that somehow individuals sacrificed themselves for the benefit of the group and he organized an array of evidence to support this view.

Wynne-Edwards' book was subjected to a detailed critique by G. C. Williams (1966), who found all of Wynne-Edwards' data wanting. In 1964 W. D. Hamilton published the first of a series of papers that attempted to reconcile the individualism of selection with the preservation of certain behaviors beneficial to the group at the expense of the individual (Hamilton 1964, 1970, 1971a, 1971b). This effort resulted in the creation of the concept of "inclusive fitness." Wilson succinctly renders it as "the sum of an individual's own fitness plus the sum of all the effects it causes to the related parts of the fitnesses of all its relatives" (Wilson 1975:118).

This seemingly simple concept accounts for the emergence of sociobiology. It argues that socially beneficial behavior can develop and be

maintained in populations by evolutionary processes already understood, without need to invoke some vague notion of group selection.
To the extent that certain behavior is beneficial to other members of
the group closely related to the individual who exhibits such behavior, acts of self-sacrifice can make evolutionary sense. So long as the
benefits that related group members derive exceed the costs to the
individual, the behavior increases fitness.

To be sure, operationalization of this set of notions is extremely
difficult. These practical problems have brought considerable refinement in the formulation of the arguments. But the fact remains that
the concept of kin selection attempts to resolve a major problem that
had blocked the application of evolutionary principles to the analysis
of social behavior. Whether or not the idea must ultimately be refor-
mulated, its importance cannot be questioned.

Application of sociobiological arguments to the study of humans,
as well as to other social species, is not some diabolical ploy, the ex-
cesses of certain practitioners notwithstanding. This important new
development in biological science is relevant to at least some social
species and it is reasonable to entertain possible applications to humans. If arguments thus far advanced in regard to humans cannot be
taken very seriously, they do not invalidate the enterprise.

No one should underestimate the harsh empirical requirements to
be met in such an analysis. We need past and present population sizes;
complete, accurate pedigrees; random mating system (unless the form
of many sociobiological propositions is changed considerably); and
typologies of "fitness-enhancing reciprocities," along with concrete
data about their reproductive effects. While all evolutionary research
involves empirical compromises that fall far short of perfection, compelling samples of data on these points are minimum requirements to
be met before anyone can say that data exist to support or disprove
the stronger sociobiological propositions as applied to humans. That
such evidence has been less the center of attention than adaptive story-
telling (pro and con) is part of the ambivalence surrounding sociobiology that needs to be unraveled.

The central question I ask of Wilson's work is whether or not he
applies any particular element of specifically sociobiological theory to
humans. The answer is no. Wilson's views on humans have not prof-

ited from the intriguing propositions sociobiology could generate. What he says about humans was said not only before sociobiology came about but before evolutionary biology as a whole. Nor is this some personal peculiarity of Wilson's thought. In conceptualizing human nature, Wilson unknowingly reproduces a pre-evolutionary view of the relationship between nature and culture, thus failing to apply evolutionary analysis to human behavior and demonstrating the pervasive power of cultural systems.

Sociobiology: The New Synthesis

Morality, Selfishness, Altruism, and Kinship

Sociobiology's first chapter, "The Morality of the Gene," begins by taking issue with Albert Camus's statement that suicide is the only important philosophical question. It is the biologist, Wilson claims,

> who is concerned with questions of physiology and evolutionary history, [who] realizes that self-knowledge is constrained and shaped by the emotional control centers in the hypothalamus and limbic system of the brain. These centers flood our consciousness with all the emotions . . . that are consulted by ethical philosophers who wish to intuit the standards of good and evil. What . . . made the hypothalamus and limbic system? They evolved by natural selection. . . . This brings us to the central theoretical problem of sociobiology: how can altruism, which by definition reduces personal fitness, possibly evolve by natural selection? The answer is kinship. [Wilson 1975:3]

Then by a leap that has not worked for anyone, Wilson implies that understanding the material structure of the brain and the evolutionary process by which it came into being creates direct understanding of the content of human thought. We can adjudicate, he implies, between particular thoughts (in this case life versus suicide) by reference to biological structures and their evolution. This theme persists in his other works as well.

While the necessary material structure of the human brain does in an ultimate sense constrain what can be thought, these constraints relate so remotely to our ability to predict the content and structure of systems of ideas that Wilson's formulation cannot be taken seriously. Almost no one who accepts evolutionary theory will dispute the point that the hypothalmic and limbic systems evolved by natural selection or that we must learn why altruism is evolutionarily possible. But this knowledge will not automatically lead us to moral clarity.

Wilson's prose suggests that selfishness and altruism exist in a pitched battle, though nothing in the theory of inclusive fitness suggests that they must. He evokes an image of humanity torn between ambivalent impulses programmed into our brains and argues that understanding the conditions that led to this impasse will permit us to control our behavior. This is an optimistic view of the human condition with strong Freudian overtones. Nothing in it is entailed in the theory of inclusive fitness.

These very first paragraphs show something about Wilson's use of words that will compound confusion later on. "Morality," "selfishness," "altruism," and "kinship" are all words that directly imply a cultural capacity for abstract thought, for deliberative behavior. Wilson's use of terms taken from the cultural world humanizes the nonhuman world by imputing morality, selfishness, altruism, and kinship to cultureless creatures. Then by reverse extrapolation he applies these terms to humans. It then appears that we are just like all the other animals. This is just linguistic sleight-of-hand, a point Marshall Sahlins (1976) has made eloquently. That we are animals no one can doubt. That we are just like any other animal is less clear. We are biocultural animals—not nobler or better, but different.

From Nature Through Mind to Culture

These early pages set the baseline for Wilson's whole argument about the relationship between genes, mind, and culture. His reductive program for eliminating the distance between culture and biology operates by rhetorical means that have little or nothing to do with sociobiological theory proper. Even Wilson does not follow his reductionism: he holds "rationality" to be above the realm of direct biological causation, while making it crucial to our species' biological salvation.

A highly social species such as man "knows," or more precisely it has been programmed to perform as if it knows, that its underlying genes will be proliferated maximally only if it orchestrates behavioral responses that bring into play an efficient mixture of personal survival, reproduction, and altruism. Consequently . . . the conscious mind [is taxed] with ambivalences whenever the organisms encounter stressful situations. [P. 4]

The mind is simply a complex apparatus that overlies the genes and must necessarily act in the interest of perpetuating the genes of that organism. The mind is a fitness-informing device. In this way Wilson drives a wedge between the genes as an ultimate level of reality and the conscious mind as an environmental tracking device that calculates fitness outcomes of various courses of action.

Chapter 2 begins in a striking way with the following lines: "Genes, like Leibnitz's monads, have no windows; the higher properties of life are emergent" (p. 7). Here is an interesting conundrum. Much of the book argues that genetic causality is the only form of "real" biological causality. Yet here a combined argument for holism and emergent levels of organization is made the centerpiece. Wilson wants to use two incompatible views of the organization of the world as they suit his convenience.

Sociobiology is the study of the biological foundations of all social behavior. As genes underlie the mind, so sociobiology supposedly underlies sociology and the humanities. "It may not be too much to say that sociology and the other social sciences, as well as the humanities, are the last branches of biology waiting to be included in the Modern Synthesis" (p. 4). This remark threatens many territories and has been widely cited. Wilson claims to recognize no general causes of behavior that are not biological, and so the incorporation of the social sciences and humanities in the sociobiological synthesis will be accomplished according to the ground rules of biology. While this idea in itself is not bad, since nothing in the social sciences and humanities could in any ultimate sense conflict with the biological capacities of human beings, Wilson's terms of incorporation destroy rather than explain the social sciences and the humanities.

Wilson defines society as "a group of individuals belonging to the same species and organized in a cooperative manner" (p. 7). This vague and analytically useless definition of society is purposely broad, Wilson says, so that it can apply to almost any aggregation of a species in which some small degree of interaction occurs. In fact, this vagueness does not seem costly at the outset, but it becomes so when the similarities and differences in the social forms of different species and ultimately in human and other animal societies are examined.

Wilson provides an elaborate discussion of various mechanisms and effects related to the rate of the evolution of social behavior. He endeavors to develop a concept of "social drift," made up of a genetic element and a "tradition" element, in analogy to genetic drift. As an example of "tradition drift," defined as behaviors learned solely as a result of social experience, he speaks of the acceptance of a new idea in a human group. His model, which is a very old one, claims that ideas compete for acceptance, and the best variant survives.[1]

We can see here the weakness of Wilson's approach to cultural analysis. The mechanistic treatment of ideas apart from their content and contexts is inexcusable. Further, the modality for communicating ideas between parent and child and among cohorts is linguistic communication, a system of transmission dependent on distinctive cultural mechanisms. The acceptance or rejection of an idea is as significantly conditioned by its fit within a larger system of ideas and by the modes in which it is communicated as by any inherent strength or weakness in the idea itself.

Though the concept of tradition drift has something to recommend it, especially as applied to nonhuman animals (whose social learning has been underemphasized by scholars until recently), the application to humans reveals important weaknesses in Wilson's thinking. He abstracts out a prime characteristic of culture, but in the process he impoverishes the concept of culture beyond recognition. It is abundantly clear that he does not use the concept "symbol," "symbolic system," "context," or "meaning" in acceptable ways. These weaknesses ultimately ruin his discussion of humans.

Speaking of group size in an evolutionary context, Wilson uses the example of the Mennonites in the rural United States as proof that mechanisms found in other animal societies work for humans. This choice is quite revealing. Not only are the demographic data adduced very weak, but he forgets that the boundaries of Mennonite communities are religiously defined. Yet the fluctuation he finds in Mennonite group size is said to represent the operation of universal mechanisms. How many macaque communities are bounded religiously and are ethnically oppressed? More important, Wilson's inclusion of the Mennonites does not enhance our understanding of them at all because it is already known that communal agricultural societies have an optimum size that varies according to changes in land base, technology, and communication. How does the use of evolutionary language improve our understanding of either the Mennonites or the macaques? If this is what Wilson means when he says that human behavior is "consistent" with sociobiological theory, then I see no difference between "weak" inferences and useless ones.

Wilson also misses opportunities to apply his models to humans well. When he discusses adjustable group size and describes societies that adjust their size to available resources, he does not mention humans. This is one subject on which there is somewhat better human evidence (Lee and De Vore, eds., 1968). It appears that cultural systems are great facilitators of the expansion and contraction of group sizes and that kinship networks (in the correct anthropological sense of the term "kinship") serve to enhance the ability of groups to fuse and divide. Here, where a human example would be worth thinking about, Wilson overlooks the opportunity.

Later, after arguing that the correct definition of higher organisms is the degree of refinement in their ability to adjust to the environment (p. 151), he takes up tradition once again. "The highest form of tradition . . . is of course human culture. *But culture aside from its involvement with language*, which is truly unique, differs from animal tradition only in degree" (p. 168; emphasis mine). This statement is quite remarkable. Wilson uses a radical distinction here between humans and animals, yet presumably a major point of the book is to moderate just such a distinction.

The phrase "culture aside from its involvement with language" is incomprehensible. Ordinarily we define culture as a congeries of symbolically mediated behaviors that have some systematic internal organization. In the social sciences and the humanities we have often considered language a major paradigm for what culture in general is like. There is also wide agreement that the development of language is the key to the development of culture; that without one the other cannot exist. And finally, the absolute uniqueness of human language itself is being questioned by the very ethologists from whom Wilson otherwise draws so much sustenance. What, then, can Wilson mean? Unfortunately, the only interpretation that can be placed on this crucial passage is that Wilson does not know what he means by either "culture" or "language."

The chapters on communication support this contention. Though Wilson tries to use language as paradigmatic for communication systems (p. 177), he rejects the universal design features of language, is confused about phonemes, and entirely forgets that language is not analyzable without reference to meaning. And then, having used language as a paradigm for communication system in general, he reverses the field and argues that human language is unique (pp. 201–2) and that the application of human language concepts to other animal communication systems is risky.

In part, Wilson's problem is simply one of expertise. The material on aggression, spacing, and dominance is better handled. He retains a lively sense of multiple causes and multiple effects and he balances predictive statements with reasonable caution. Here he is clearly on familiar ground. Except for one careless aside on "obvious parallels" with humans in his discussion of the will to power (p. 287), this set of chapters, in which all sorts of bits and pieces of human evidence are fitted in, is not marred by the kind of outlandish comparisons that came earlier. Knowing this material better, Wilson is more diffident about extrapolation.

After a discussion of what he terms "role" and "caste" among nonhuman animals, Wilson turns to human roles.

> But whereas, social organization in the insect colonies depends on programmed, altruistic behavior by an ergonomically optimal mix of castes, the welfare of human societies is based on trade-offs among individuals playing roles. When too many human beings enter one occupation, their personal cost-to-benefit ratios rise, and some individuals transfer to less crowded fields for selfish reasons. [P. 313]

This statement could have been made by Galton, Malthus, or Milton Friedman. It reveals a naive free-competition model of society without any awareness of problems of social stratification and power or of the long history of debate on this subject.

Species Immortality

Since modification of the environment is a particularly marked human characteristic, Wilson's comments on this subject have considerable importance.

> Manipulation of the physical environment is the ultimate adaptation. If it were somehow brought to perfection, environmental control would insure *the indefinite survival of the species*, because the genetic structure could at last be matched precisely to favorable conditions and freed from the capricious emergencies that endanger its survival. No species has approached to environmental control, not even man. [Pp. 59–60; emphasis mine]

This statement needs to be remembered, for it contains the core of Wilson's peculiar utopianism. If we humans could manipulate the environment rather than letting it affect us, we could become our own ultimate causes in the world. And were we so inclined—as Wilson seems to be—we could try to bring evolution as we know it to a halt. This goal is nothing less than the achievement of species immortality in the material world. It is Wilson's alternative to the immortality of the individual soul.

Scholarly emphasis on human manipulation of the physical environment has led many authors to argue that culture has taken over from biology among humans. Wilson does not agree. He calls such manipulation an "adaptation," thus insisting that culture be treated as one more biological adaptation. While this position is generally reasonable, Wilson finds it necessary to ignore the symbolic and systemic aspects of culture.

Why does Wilson consistently ignore these aspects? I believe it is because he sees culture (in contrast with science) as obscuring our view of truth. Only when we purge culture of its irrational elements will culture give a true picture of the environment. Then we can reach the ultimate adaptation. Wilson thus wishes to reduce culture to its scientific-rational components. The rest of culture must be consigned to the dustbin.

"Man: From Sociobiology to Sociology"

Wilson has been repeatedly drubbed for his final chapter. Indeed, he published *On Human Nature* to remedy just this problem. But the errors in this chapter merit comment because they help to reveal the major cultural presuppositions that underlie Wilson's failure to apply evolutionary analysis to humans. All the difficulties discussed earlier now combine and interact.

The chapter begins with an invocation of an extraterrestrial zoologist, presumably because from an extraterrestrial perspective we humans could not deny that we are animals. Noting that we are ecologically "peculiar" because we are so wide ranging and locally dense in some areas, Wilson also stresses our anatomical uniqueness.

> We have leaped forward in mental evolution in a way that continues to defy self-analysis. The *mental hypertrophy* has distorted even the most basic primate social qualities into nearly unrecognizable forms. Individual species of Old World monkeys and apes have notably plastic social organizations; man has extended the trend into a protean ethnicity. Monkeys and apes utilize behavioral scaling to adjust aggressive and sexual interactions; in man the scales have become multidimensional, culturally adjustable, and almost endlessly subtle. Bonding and the practices of reciprocal altruism are rudimentary in other primates; man has expanded them into great networks where individuals consciously alter roles from hour to hour as if changing masks. It is the task of comparative sociobiology to trace these and other human qualities as closely as possible back through time. Besides adding perspective and perhaps offering some sense of philosophical ease, the exercise will help to identify the behaviors and rules by which individual human beings increase their Darwinian fitness through the manipulation of society. In a phrase, we are searching for the *human biogram.* . . . One of the key questions . . . is to what extent the biogram represents an adaptation of modern cultural life and to what extent it is a phylogenetic vestige. Our civilizations were jerrybuilt around the biogram. How have they been influenced by it? Conversely, how much flexibility is there in the biogram, and in which parameters particularly? Experience with other animals indicates that when organs are hypertrophied, phy-

logeny is hard to reconstruct. This is the crux of the problem of the
evolutionary analysis of human behavior. [P. 548; emphases mine]

There is much to consider here. "Hypertrophy" suggests an almost
unnatural overgrowth of an organ. By what standards do we judge
this condition? Do birds have hypertrophied digits? What is the difference between hypertrophy and a complex morphological adaptation? This is really an issue in classification. Wilson uses the term to
suggest that humans may have gone too far in one direction and that
we are much in need of perspective and self-control. As a biological
concept in regard to humans, mental hypertrophy is vacuous.

At the same time that Wilson evokes human variability in ethnicity
and in social roles, he darkly invokes a biogram that must necessarily
set limits around the protean character of humanity. Biology teaches
us, he suggests, what these limits are so we can know how to behave.

Surely this make no sense in view of his general theory. If, as he has
insisted throughout the book, he is a biological determinist, then humans cannot behave in any way that is not biologically feasible. If
this is the case, what is there to worry about? But Wilson is obviously
worried. The true meaning of mental hypertrophy becomes clearer
now. He thinks that overdevelopment of the brain can lead us to think
and behave in ways that are not consistent with our survival. If we
want to survive as a species, we must come back to reality and analyze the true evolutionary constraints that affect us.

Clearly this is a peculiar problem for an evolutionist to worry about.
No other species concerns itself with species immortality. Species adapt
or not; they continue or become extinct. For all his emphasis on evolution, Wilson finds such a fate intolerable for humanity. We should
try to develop the perfect adaptation and become immortal as a species. While the appeal of this view is understandable, it has no conceivable connection to sociobiology and is only tenuously related to
evolutionary biology. It is also a view that such thinkers as Malthus,
Galton, Lorenz, and Desmond Morris have held without reference to
sociobiological theory at all.

After this strange beginning, Wilson deals with human flexibility in
more detail. He speaks of "ecological release" through lack of competition with other species (p. 550), and he christens the human ca-

pacity for flexible behavior genetic "underprescription" (p. 559). This amounts to a double renaming of what most anthropologists would simply call cultural behavior.

Rudimentary discussions of language, the nuclear family, and other issues show how far out of his own area of expertise Wilson has strayed. These divagations should not be taken too seriously because the real point comes when Wilson tries to sharpen his analysis of culture.

"Culture, including the more resplendent manifestations of ritual and religion, can be interpreted as a hierarchical system of environmental tracking devices" (p. 560). Culture change and environmental change thus occur at similar rates. Religion, however, apparently interferes with such tracking:

> Formal religion . . . has many elements of magic but is focused on deeper, more tribally oriented beliefs. The enduring paradox of religion is that so much of its substance is demonstrably false, yet it remains a driving force in all societies. Men would rather believe than know, have the void as purpose, as Nietzsche said, than be void of purpose. . . . The individual is prepared by the sacred rituals for supreme effort and self-sacrifice. . . . *Deus vult* was the rallying cry of the First Crusade. God wills it, but the summed Darwinian fitness of the tribe was the ultimate if unrecognized beneficiary. [P. 561]

This situation is apparently connected to hypertrophy. Definable evolutionary conditions have led us to mental hypertrophy, which has increased our capacity for flexible behavior. But this flexibility is now hedged round by the irrationalism of religion, which has monopolized the means of indoctrinating people with regard to altruistic behavior. The "demonstrably false" religions are taking our hypertrophy and turning it into a danger for our species. This danger must be met, and the answer is sociobiology:

> It seems that our autocatalytic social evolution has locked us onto a particular course which the early hominids still within us may not welcome. *To maintain the species indefinitely* we are compelled to *drive toward total knowledge*, right down to the level of the neuron and gene. When we have progressed enough to explain ourselves in these mechanistic terms, and the social sciences come to full flower, the result might be hard to accept. . . . But we still have another hundred years. [P. 575; emphases mine]

"To maintain the species indefinitely . . . total knowledge." This is Wilson's true agenda. Our species should strive to maintain itself indefinitely by learning scientifically how evolution applies to us. Thus we must push aside religion and the other "cultural mystifications" that hide what we really are.

This is really an old call to the imposition of rational science over irrational religion on the promise of a utopian future. In this utopia the antithesis between nature and culture will have been abolished by science. The argument assumes that all that is truly human and worthwhile is rational, and that science is thus the quintessence of humanism (as against the false humanism of the so-called humanities). Surely this is the antithesis of the scientific method.

Scientific Method and Loose Thinking

One way of dealing with some of the most patent inconsistencies in *Sociobiology* is to claim that Wilson is simply a bad scientist, or at least a very naive one. Such a convenient view does not account for the data and is much too easy a way out of a complex problem. Wilson is a famous and widely respected scientist who clearly understands the canons of scientific method. His abstract discussion of the theoretical structure and requirements of sociobiology demonstrates this understanding. He emphasizes a distinction between ultimate and proximate causality, one that is now much bandied about. By "proximate causation" he means essentially such immediately functional causes as anatomy, physiology, and behavior. By "ultimate causation" he means the necessities created by the environment (p. 23). Clearly some such distinction is important in most dynamic analyses and in analyses where differences in scale are important. Yet these distinctions, unless they are carefully handled, are a perfect escape clause that can protect a theory from empirical challenge. If inconvenient evidence is found at one level, then causality at the other level can be invoked, and Wilson does invoke it repeatedly in his human examples.

In the section "Reasoning in Sociobiology" Wilson gives a fair characterization of the deductive basis of science. He discusses the use of "strong inference" and criticizes the "advocacy" method of proof.

Arguing in favor of multicausal theories in sociobiology that move the various levels of analysis together in a sensible way, he concludes, "The goal of investigation should not be to advocate the simplest explanation, but rather to enumerate all of the possible explanations, improbable as well as likely, and then to devise tests to eliminate some of them" (p. 30). One can only agree. This is a textbook scientific method. That Wilson is aware of these rules is important because, as we shall see, the requirement to develop a variety of hypotheses, devise tests, and apply them is dropped when his subject is humans.

By Chapter 5 he has moved far from these elegant statements about scientific method. Such comments as the following are found:

> Human behavior abounds with reciprocal altruism consistent with genetic theory. . . . The critical gene frequency is simply that in which playing the game pays by virtue of a high enough probability of contacting another cooperator. The machinery for bringing the gene frequency up to the critical value must lie outside the game itself. It could be genetic drift in small populations . . . or a concomitant of interdemic or kin selection favoring other aspects of altruism displayed by the cooperator genotypes. [Pp. 120–21]

The method of strong inference is gone, and with it the elaboration of multiple hypotheses and the use of tests to eliminate some. Proofs regarding human behavior in particular hang on the words "consistent with," a slippery phrase that says "caused by" without really defending or testing the proposition.

Wilson concludes Part I with the following statement:

> Although the theory of group selection is still rudimentary, it has already provided insights into some of the least understood and most disturbing qualities of social behavior. Above all, it predicts ambivalence as a way of life in social creatures. . . . In the opening chapter of this book, I suggested that a science of sociobiology, if coupled with neurophysiology, might transform the insights of ancient religions into a precise account of the evolutionary origin of ethics and hence explain the reasons why we make certain moral choices instead of others at particular times. Whether such understanding will then produce the Rule remains to be seen. For the moment, perhaps it is enough to establish that a *single strong thread* does indeed run from the conduct of

termite colonies and turkey brotherhoods to the social behavior of man.
[P. 129; emphasis mine]

Wilson promises that sociobiology can convert religion and morality into science by reducing them to evolutionary theory. This moral-
ized science promises to save humanity by purging culture of its irrational elements and bringing us into concert with the environment
through reason. The scientific method and strong inference have been
supplanted by the advocacy method. That a "single . . . thread" runs
from termites to man could be true in any typological system (e.g.,
we both locomote by means of limbs of some sort). The strength of
his thread is supplied by the logic of the system he has created, not by
any tests he devised or applied. The thread is strong only if we already
believe Wilson.

On Human Nature

In writing *On Human Nature* Wilson had much damage to repair.
The book is a great disappointment in this regard. It does, however,
confirm my reading of the cultural system that underlies the views
expressed in *Sociobiology*.

On Human Nature is an avowedly speculative view of the union of
the natural and the social sciences. We shall see that the terms of
union make it an annexation of the latter by the former. Not as compellingly organized as *Sociobiology*, it adopts a topical approach to
aspects of human nature encapsulated in such chapter titles as "Dilemma" and "Hope." The body of the work contains a disappointing
array of observations about humans.

Since the principles of both sociobiology in particular and evolutionary biology in general are suspended in the chapter on humans in
Sociobiology, one approaches *On Human Nature* with hope that this
failing will have been at least partly rectified. It has not. Indeed, its
contents are indistinguishable from those of popular works by such
authors as Lorenz, Ardrey, and Morris.

Sociobiology, given its important new formulations regarding the
evolution of social behavior, should make some notable changes at

least in the phrasing of evolutionary questions about human behavior. Yet one seeks in vain for new perspectives on human behavior in *On Human Nature*. The deviations from evolutionary analysis effectively domesticate sociobiology so that it preserves the traditional Western view of human nature while covering it with the terminological trappings of Darwinism.

"Man's Ultimate Nature," Natural Reason, and Truth

Chapter 1 opens with this question: "What is man's ultimate nature?" The question itself betrays the fundamental orientation of the work. What can the "ultimate nature" of a species mean if a species is a congeries of ranges of variation that are continuously shifting? Can we talk about the "ultimate nature" of a species in evolutionary biology? Certainly not. This concept only fits the chain-of-being model of creation. Thus the book begins on a nonevolutionary note.

This question is immediately followed by a statement that distracts our attention from the issues just raised and focuses on the specter of materialism: "If the brain is a machine of ten billion nerve cells and the mind can somehow be explained as the summed activity of a finite number of chemical and electrical reactions, boundaries limit the human prospect—we are biological and our souls cannot fly free" (Wilson 1978:1). This statement introduces a theme that runs throughout the work. Biology is a constraint on culture. To be realistic we must adjust our culture to this fact. Such a view of the relationship between biology and culture is inappropriate to modern biology, but it is common in the works of Hippocrates, Jean Bodin, and other pre-Darwinian writers.

Wilson has a strong tendency to link such concepts as soul with religion and culture, and to link such concepts as science and rationality with transcendence of the limits of culture. Wilson naturalizes reason: "The human mind is a device for survival and reproduction, and reason is just one of its various techniques"; "Human nature can be laid open as an object of fully empirical research, biology can be put to the service of liberal education, and our self-conception can be enormously and truthfully enriched" (p. 2). Reason is natural; natural science is about what is natural; what is natural is real and true.

Thus natural science can tell us the truth about ourselves and move our reasoning onto a mature plane, far from the fantasy world of religion and the humanities.

> In order to search for a new morality based upon a more truthful definition of man, it is necessary to look inward, to dissect the machinery of the mind and to retrace its evolutionary history. But that effort, I predict, will uncover a second dilemma, which is the choice that must be made among the ethical premises inherent in man's biological nature. [Pp. 4–5]

Biological Constraints and Moral Choice

According to *Sociobiology*, we are programmed for both selfishness and altruism. The only moral choice Wilson can understand is between these alternatives. In *On Human Nature* he eliminates this moral choice by arguing that it is rational to be altruistic. Further, he wants to use biology to prove that after we have measured the tightness of the material constraints, we still have the freedom to choose.

> The challenge to science is to measure the tightness of the constraints caused by the programming, to find their source in the brain, and to decode their significance through the reconstruction of the evolutionary history of the mind. . . . [We will then be able to decide which] of the censors and motivators should be obeyed and which one might better be curtailed or sublimated. . . . [P. 6]

This is an odd position for an author who begins a book with an epigraph from Hume, who compellingly portrays the distance separating "is" from "ought."

Wilson really only plays with the idea of constraint. He pretends to measure constraint to prove that we have the freedom to choose, that rationality has a role to play. Yet by annihilating the understanding of cultural systems in both of his books, he deprives that appeal to rationality of any context or meaning.

The question of constraint comes up again in a variety of forms: "The question is no longer whether human social behavior is genetically determined; it is to what extent" (p. 19). It is a serious error to attempt to analyze the relationship between biology and culture as a single continuum ranging from fully biological to fully cultural and then to place traits along the continuum. As Lewontin ([1974] 1976) has argued more powerfully than anyone else, this is bad biology.

All culture is biological, for without biological beings there is no culture. But if we agree that all culture is 100 percent biological in this sense, we have said nothing useful about constraints, freedom, culture, or behavior. Wilson has simply restated the old dichotomies—environment/culture, nature/nurture, genes/culture, constraint/freedom. These polarities do not belong in evolutionary biology.

Wilson confuses the issue further: "Either possibility—complete cultural determination versus shared cultural and genetic determination of variability within the species—is compatible with the more general sociobiological view of human nature" (pp. 42–43). Having started from the position that everything humans do is biological and material, he here argues that some things may be cultural without being biological. This confusion is nothing more than an expression of the old dualistic model of human nature.

Wilson's quest for the ultimate nature of humans also leads him into trouble with biological diversity. He turns the problem over and over and finally tries the following formulation: "Hope and pride and not despair are the ultimate legacy of genetic diversity, because we are a single species, not two or more. . . . Mankind viewed over many generations shares a single human nature. . . ." (p. 50). Whatever this statement means, and I challenge others to make sense of it, it only highlights the problem of trying to assert a species essence in the humoral/environmental sense and biological diversity in an evolutionary biological sense. The positions are irreconcilable.

At the end of the work, Wilson returns to the question of moral choice. Given our ambivalent "essence," we are biologically programmed to be free to choose between selfishness and altruism. But in his view, science tells us that only altruism is rational. As he puts it, "circularity of the human predicament is not so tight that it cannot be broken through an exercise of will." Sociobiology "will fashion a biology of ethics, which will make possible the selection of a more deeply understood and enduring code of moral values" (p. 196). Then he provides this biology of ethics:

> Because natural selection has acted on the behavior of individuals who
> benefit themselves and their immediate relatives, human nature bends

us to the imperatives of selfishness and tribalism. But a more detached view of the long-range course of evolution should allow us to see beyond the blind decision-making process of natural selection and to envision the history and future of our own genes against the background of the entire human species. A word already in use intuitively defines this view: nobility. Had the dinosaurs grasped the concept they might have survived. They might have been us. [P. 197]

This remarkable passage sums up Wilson's true agenda. The mental hypertrophy that characterizes humans should allow us to see past selfishness as shortsighted and to realize that altruism, though in the short term possibly disadvantageous, in the long term will ensure our survival. Evolution is blind but it has produced a creature capable of vision. Our will and our reason can permit us to outsmart the environment and approach the ideal of bringing evolution to a halt. We must look to empirical biological research for our ethical systems; we must derive "ought" directly from "is." The price for failure is ongoing evolution, which may leave us as extinct as the dinosaurs.

Evolutionary "Truth"

In Wilson's view, what most encumbers our vision is the "falsity" of our cultural systems. Our religious ideas are especially at fault because they supposedly deny the materiality of human life and celebrate the irrational. To address this problem Wilson suggests that we drop the "biblical epic" and put the "evolutionary epic" in its place.

The core of scientific materialism is the evolutionary epic. . . . What I am suggesting . . . is that the evolutionary epic is probably the best myth we will ever have. It can be adjusted until it comes as close to truth as the human mind is constructed to judge the truth. And if that is the case, the mythopoeic requirements of the mind must somehow be met by scientific materialism so as to reinvest our superb energies. [P. 201]

Man's destiny is to know, if only because societies with knowledge culturally dominate societies that lack it. [P. 207]

The problem is that our cultural knowledge thus far in human history has been "false." Now we can make it "true."

What is the truth? It appears that the truth is that a judicious combination of selfishness and altruism is the only evolutionary strategy that will work. This is universally true no matter what the circumstances or time period. Under these conditions, the only moral thing to do is to follow the dictates of scientific reasoning, which has uncovered the universal and eternally best strategy for survival. Thus Wilson ends by spiritualizing biological science, trying to convert its results into direct guidelines for behavior based on the analysis of the diversity of species now existing and their evolutionary histories.

Genes, Mind, and Culture

In collaboration with the physicist Charles Lumsden, Wilson has made yet another attempt to deal with humans in a way that is supposed to be sociobiological. To Wilson's credit, he not only keeps trying to strengthen his position but also recognizes basically where the difficulties in the enterprise lie. The relationship of the genetic, mental, and cultural components of human behavior has been the central difficulty, and *Genes, Mind, and Culture* (Lumsden and Wilson 1981) deals directly with this problem.

As in the case of *On Human Nature*, one reads this book in the hope that the enormous amount of criticism leveled at Wilson's two earlier works will have significantly sharpened his formulation. Despite an improved lexicon, a complex statistical apparatus, and wider reading in cultural anthropology, Wilson has become so entangled in the difficulties already described that he has moved away from rather than toward his goal.

In the Preface Lumsden and Wilson argue that genetic and cultural evolution must be linked, and that the connection between them may be found in "the ontogenetic development of mental activity and behavior" (Lumsden and Wilson 1981:ix). This argument is coupled with a criticism of sociobiology for having failed to deal successfully with the operations of the human mind and with the immense amount of cultural diversity found even in the contemporary human world. This is an encouraging beginning.

The Introduction contains virtually unexceptionable statements about the necessary relationship between genes and culture: "We view it . . . as a largely unknown evolutionary process—a complicated, fascinating interaction in which culture is generated and shaped by biological imperatives while biological traits are simultaneously altered by genetic evolution in response to cultural innovation" (p. 1).

Except for the use of the indefensible term "cultural evolution" (a fault the authors share with most other practitioners in this field),[2] this reasoning makes sense. No doubt there must be a relationship between genes and culture. Conceptualizing it as a complex, interactive process seems currently the most promising way to move the discussion. Yet within a few pages the authors manage to extinguish all enthusiasm for their approach by returning to the contradictions that flawed Wilson's previous works.

Early on we are faced with the possibility of "pure cultural evolution," which is somehow dependent on genetic controls. The conceptualization of the relationship between genes and culture is as confused as ever. Once again Wilson views the relationship between genes and culture as one in which genes hold culture "on a leash" (now called the "leash principle" [p. 13]). Most of the book is devoted to mathematical estimates of the varying length of the leash.

Without genes there is no culture. Yet genes hold culture on a leash. The incompatibility of these two views is clear. If we argue, as we must, that without genes there is no culture, then the only reasonable idiom for discussing the relationship between genes and culture is one involving the analysis of the relationship between levels of causation, perhaps in the general systems mode. But if we follow this course, it makes no sense to say that genes hold culture on a leash. The leash principle asserts that there is one entity in the world to be called "genes" and another to be called "culture," and that the degree to which the former ordains the details of the latter is determined by the length of the leash between them. The leash principle reproduces the old nature/culture argument while the idea that there is no culture without genes demolishes the distinction.

Because this approach is deeply wound into the internal structure of the work, it compromises the rest of *Genes, Mind, and Culture*. Culture is treated as an adaptive system without reference to the problem of meaning. The mind remains a fitness-informing device concerned with survival and reproduction; and evolution is again treated as a myth that is powerful only because it is "true."

Wilson still recoils from the notion that the apparent complexity of human sociocultural arrangements must be taken seriously rather than reduced to a few homilistic rules. "Although *Homo sapiens* is the most complex species on earth by a spectacular margin, it is probably far less complex and difficult to understand than contemporary social theory leads one to believe" (p. 350).

He then takes aim at hermeneutics as a kind of obscurantist philosophy designed to descend into an antiscientific universe of cultural details. He tips his hat to Marx's attempt to formulate the laws of history, saying that the effort was a good one that now can be carried forward with the help of sociobiology. And the book closes roughly where *Sociobiology* closed, all the evocations of interactive biocultural processes and of cultural complexity having produced no substantial effect. No more convincing display of the power of culture to invoke and maintain meanings can be found than Wilson's own derailment of evolutionary biology in the service of his nonevolutionary view of the relationship between nature and culture.

Wilson as an Evolutionist

It is clear that when Wilson deals with humans, he uses neither sociobiology nor evolutionism. Rather he is a naturalistic thinker in the tradition of those whose ideas rest on a dichotomous static vision of nature and culture. Except for the occasional loose connection made between kin selection, reciprocal altruism, and cooperative behavior, Wilson is content to use biological information to make the case that we humans are in grave danger of extinguishing ourselves because of the contradictions in our "nature."

Most of the major rules of evolutionary biology are overturned. Despite some remarks here and there about diversity in human populations, Wilson speaks of humans as having a "nature," indeed as having an "ultimate nature" that is potentially permanent. Nothing in evolutionism permits such assertions. A species is an interbreeding congeries of varying traits, continually in motion, with fuzzy boundaries in fact, if not by definition.

Wilson's goal is to preserve our species from extinction by perfecting our ecological release via proper balancing of our selfish and altruistic natures. While the goal of preservation of our species may be laudable, nothing in evolutionary biology justifies such actions. Evolutionary biology is not about preserving species indefinitely; it is about the mechanisms and trajectories of evolutionary change.

In the most bald fashion, Wilson repeatedly moves from what he considers to be the facts of evolutionary biology to the moral imperatives he thinks we must follow. He leaps across the boundary from "is" to "ought" without regard for most thinkers' belief that such a leap is rationally unjustifiable and ideologically compromised. He thinks that a philosophy of empiricism will simply and directly solve our moral dilemmas.

This belief finds no support in evolutionary biology. Evolution is a theory, not a description of neutral facts. Evolutionary theory produces no clear moral imperatives. We may weigh the evolutionary consequences that can arise from various courses of action, but the choice of a course of action and the imposition of that choice on others in our society cannot be justified on simple evolutionary grounds.

Wilson's pattern of thought is not explicable in terms of evolutionary theory but it does fit the old Western view of the human epic. Humans began in a state of nature in which natural laws virtually controlled all. As we became increasingly successful and proliferated, our cultural ideas became more and more complex, obscuring the "real" (natural) requirements of human existence. We began to act in ways not in our collective interest. But this dangerous cultural world also gave rise to the scientific method, and its advance ushered in the possiblity of a final age of human evolution in which scientific purging of untrue cultural ideas will permit us self-consciously (rationally) to regulate our relationships with nature, using reason to abolish the contradiction between nature and culture. This is an elite-managed rational utopia.

While a knowledge of evolutionary biology does not help us understand what Wilson thinks, an understanding of medieval theories about the relationship between purity of blood and social order does. The genealogical principle dominates in those theories, and blood "holds culture on a leash"; the social order is a reflection of the biological characteristics of the population. Environmental and historical effects, such as invasions and migrations, have confused the relations between blood and social position. As a result, a scientific analytical effort is necessary for effective political management and responsible moral judgment.

Social order depends on obedience to these basic "natural" principles. "Science" served this medieval enterprise in the form of elaborate genealogical investigations to determine the "natural" constraints involved. Politics and ethics were felt to be the direct expressions of rational inquiry. History was viewed as the playing out of these natural principles. Once the principles were understood, history became an intelligible, manageable process.

The claim that Wilson's construal of the relationship between nature and culture has much in common with arguments about purity of blood is not made with the intention of ridiculing Wilson. He is a sincere and serious scholar and deserves to be treated as such. Rather it emphasizes the fundamental similarity between his views and those pre-evolutionary views in reifying the distinction between nature and culture as a way of harnessing science to political and moral judgment.

What then is sociobiological or evolutionary about Wilson's view of humans? Nothing. Wilson's human sociobiology has been fully domesticated by an ancient Western cultural vision. That his own view of human nature cannot explain analytically the power and permanence of the very cultural visions of which he is captive is the strongest reason to reject what he says.

Lest it be thought that Wilson is singled out for particular abuse, I must point out that Wilson's opponents often argue in ways that differ little from his. As is commonplace in attacks on biological determinist views such as Wilson's, some opponents take such extreme counterpositions that they end up arguing from supposed "facts of nature" to their own particular political and ethical preferences.

One of many egregious examples in the antisociobiological literature is the Sociobiology Study Group's "Sociobiology: A New Biological Determinism" (1977). After offering theoretical and empirical criticisms of Wilson's views—all perfectly legitimate and powerful—they argue that there can be no evidence for sociobiology because "*the truth is* that the individual's social activity is to be understood only by first understanding social institutions . . . we know of no relevent [sic] constraints placed on social processes by human biology" (p. 148; emphasis mine). This mode of argument mirrors Wilson's advocacy view. The Sociobiology Study Group's "scientific facts" about humans are turned into direct support for their own political and ethical scheme. This logic is as unacceptable in the Sociobiology Study Group as it is in Wilson. Both sides are guilty of attempting to create the illusion that science supports their politics and ethics.

Does this mean that human sociobiology is an enterprise devoid of merit? The very criticisms I have made force rejection of this claim. Since Wilson's approach to humans is nonevolutionary, a rejection of Wilson does not entitle us to reject human sociobiology. Whatever the possibilities of human sociobiology are, Wilson does not explore them in an evolutionary way. Thus my argument is not about the truth or falsity of human sociobiology; it is an explanation of the cultural forces at work that have led a major evolutionary biologist to reject much of the evolutionary paradigm when he deals with humans, that have made many of his critics equally antievolutionary, and that have led both to resemble medieval thinkers theorizing about the relationship between blood and social order. The genealogical principle, the environmental principle, and the static categories of species in the world, albeit in modified form, are still powerfully present, along with their political and moral accompaniments.

CHAPTER 7

Cultural Materialism

It is tempting to dismiss Wilson's reliance on theoretical and ideological structures familiar to us from pre- and nonevolutionary thought as a peculiarity of his own thinking. But Wilson's difficulties are shared in the human sociobiological writings of Richard Alexander (1979), David Barash (1979), and Richard Dawkins (1976), among others. The problem, of course, could be a peculiarity of sociobiology, irrelevant to other attempts to apply evolutionary theory to the study of humans. An analysis of Marvin Harris' "cultural materialism," however, shows the problem to be much broader. In Harris' presumed application of evolutionary ecology's energy-flow analysis to certain human cases, the same flaws are evident.

Sociobiology and evolutionary ecology are quite distinctive theoretical structures within evolutionary biology. Though they must ultimately be reconciled under the overall principles of evolutionary biology, they focus attention on quite different aspects of biological systems and use very divergent methodologies. Were cultural materialism and human sociobiology genuine applications of these theories and methods, they should be quite distinctive. Yet Harris' and Wilson's works share many of the same defects. Their similar difficulties arise not from evolutionary biology but from both authors' reproduction of key elements of pre-evolutionary naturalistic views of society in their work, in particular their common commitments to static "natural" categories and the nature/nurture dichotomy, and to deriving political and moral conclusions from the study of "nature."

Cultural Materialism

The application of frameworks based on the analysis of population/ resource relationships to human situations antedates modern evolutionary biology. It is conventional, though not therefore correct, to claim that Malthus made the first scientific statement of this approach in 1798. The recent popularity of population/resource models and theories dates from the 1960s and is associated with Paul Erlich, Barry Commoner, Garret Hardin, "Earth Day," the Club of Rome—that is, the ecology movement. Spokespersons for this perspective share the common aim of using ecological analysis as a basis for the formulation of social policies.

Despite some intriguing proposals, none of these scholars and groups really came to terms with the cross-cultural application of ecological perspectives. Certainly none of them dealt effectively with cultural diversity or the full sweep of human history. Into this gap stepped the anthropologist Marvin Harris.

Harris began these efforts in 1974 by reanalyzing some of the major problems in the interpretation of human diversity and history with the aid of ecological perspectives. He calls his approach "cultural materialism" to set it off from other ecological approaches. Cultural materialism first was brought to the attention of the general public in two very popular books—*Cows, Pigs, Wars, and Witches* ([1974] 1975) and *Cannibals and Kings* (1977)—which projected Harris into the arena of both academic and public debate.

Harris, who does not shrink from criticizing opponents, has polarized his sympathizers and detractors and now stands at the center of a series of polemics that are always revealing, occasionally entertaining, and often fruitless. Indeed, because of this polemical atmosphere, people who analyze the role of ecological analysis in anthropology are tempted not to deal with Harris. His treatment of critics is both unpleasant and unproductive. Neither he nor his supporters have shown much willingness to learn from criticism.

Harris is nevertheless important. Most generally unbiased readers of his popular books are readily convinced by his arguments and react with considerable hostility to criticism of his work, seeing criticism as an attack on science, rationality, and the experimental method. Harris' detractors, on the other side, are always mystified by the power of his seemingly weak arguments.

For this reason Harris must be taken seriously. He has an uncanny ability to capture the imagination and scientific optimism of an audience through arguments for the application of ecological principles to the analysis of human problems. Yet Harris violates evolutionary reasoning at every turn, shows little awareness of the evolutionary ecology that he claims to employ, and appeals to his audience mainly through just-so stories about adaptation, rationality, science, and democracy.

Harris returns to important aspects of the pre-evolutionary Western view of the relationship between nature and culture, and he does so by violating most of the principles of science, rational argument, and the experimental method to which he appeals. Anyone who wants to see useful applications of a balanced combination of evolutionary biology and cultural analysis to human beings has first to deal with Harris' claims.

Harris' history as an anthropologist extends back to monographs on race relations in Brazil and in Mozambique. The part of his work relevant to the persistence of nonevolutionary views in the study of humans begins with the publication of *The Nature of Cultural Things* in 1964. This book is a methodological essay on the "objective" observation of human behavior. It was followed in 1968 by his monumental *Rise of Anthropological Theory*, in which he rewrote the history of anthropology to legitimate the development of what he came to call "cultural materialism." In 1971 he published a very successful introductory textbook, *Culture, Man, and Nature,* in which the principles of cultural materialism were put into play for teaching purposes. Revised versions of the book are still in print. These works were followed by the four books I shall examine here: *Cows, Pigs, Wars, and Witches*; *Cannibals and Kings*; *Cultural Materialism*; and *America Now*. The first of these books is his attempt to explain strange cultural behaviors by cultural-materialist means; the second endeavors to make the cultural-materialist model dynamic; the third is a theoretical defense of cultural materialism and an all-out attack on its opponents; and the fourth is an interpretation of the ills of American society.

Cows, Pigs, Wars, and Witches

Originally published in 1974, *Cows, Pigs, Wars, and Witches* is a gem that rewards close reading. Tightly structured, lively, and occasionally both eloquent and convincing, it reveals a pattern of thinking that remains consistent throughout Harris' subsequent works. Here Harris throws down the gauntlet, claiming that he can explain with his method what the rest of us recognize as problems but are unable to deal with. He asserts that science is on his side, a bold claim in a bold book.

Objectivity and Science vs. "Cultural Dreamwork"

Harris opens with the statement that theory must reflect the "real" world (p. vii). Though hardly revolutionary, it immediately raises the question of what kind of "real" world he believes is out there. It will become clear that his theoretical preferences favor a view of the real world as a realm that operates in accordance with a very small number of "natural" laws. The real world for Marvin Harris is a world of energy flows. Calories and the ecosystems through which they flow are the material reality of human life to which he believes a true cultural science must refer.

Harris evokes the antiscience ideology of the 1960s, specifically the arguments that rationality in general and science in particular are the prime causes of our social problems. Whenever this antiscience specter is raised, Harris waxes furious and occasionally compelling. A staunch defender of science, Harris argues that whatever the causes of our problems are, too much scientific understanding of the causes of social life is not one of them. One can only agree with him in general and with his assertion that much more study of the material organization of the human world is vitally necessary.

But Harris' own argument goes far beyond the demand for more scientific work. He insists that the testing ground for a science of human behavior must be a coherent scientific (and therefore materialistic) explanation of the highly various, puzzling lifestyles that anthropologists have often portrayed so engagingly. The subtitle of this book, *The Riddles of Culture*, refers to cultural practices that have generally defied anthropological analysis (according to Harris). He claims to resolve these analytical problems with his theory, thus giving a scientific explanation to culture.

Despite the scope of the task, his argument is very simple: "I shall show that even the most bizarre-seeming beliefs and practices turn out on closer inspection to be based on ordinary, banal, one might say 'vulgar' conditions, needs, and activities" (p. 5). The problem is that

> practical life wears many *disguises*. Each lifestyle comes *wrapped in myths and legends* that draw attention to impractical or supernatural conditions. These wrappings give people a *social identity and a sense of social purpose*, but they conceal the *naked truths of social life*. *Deceptions* about the mundane causes of culture weigh upon ordinary consciousness like *layered sheets of lead*. [Harris [1974] 1975:5; emphases mine]

The call is to strip off the disguises, to reveal our self-deceptions by means of science.

Certain words and turns of phrase are most informative. Practical life is "disguised" in a costume created by a "wrapping" of myth and legend. The meanings people create, then, are external to the major causes of human existence. These disguises provide interpretations of experiences, but they are at base "deceptions" that cover the "naked truths of social life." In a few lines Harris has separated our material life as humans from the meanings we attribute to it. He has equated cultural interpretations of the "naked truths" with disguises and deceptions. Cultural systems of meaning are for him intrinsically false "layered sheets of lead."

The necessary implication is that Harris' own consciousness is in some sense not culturally mediated; that it is not "ordinary." If he can see the "naked truths" for what they are, his sources of conceptual objectivity must be uniquely noncultural. This is, in effect, his definition of science—the use of nonculturally mediated objectivity to strip the disguise from the naked truths of social life. His views are remarkably like E. O. Wilson's.

Harris takes up recognized major problems of anthropological interpretation to "test" his approach and to "prove" its validity. He deals with Hindu sacred cows, Melanesian pig veneration, Middle Eastern pork prohibitions, primitive war, male aggression, the potlatch, cargo cults, religious movements, and European witchcraft. Each case is treated as a challenge, a riddle to be solved.

Harris seeks to ferret out the underlying material causes of these apparently irrational phenomena to find the unitary truth behind the disguise. He is generally a pleasure to read as he goes about his task. The analysis is always interesting and occasionally genuinely provocative. There is no lack of intellectual derring-do and fun. Harris must be read to be appreciated.

Taboos against Temptation as Inferior Science

Harris' type case is the sacred cow of Hinduism. "Sacred cow" has become a Western cover term to refer to all that is irrational in cultures different from our own. Harris argues that Hindu cow veneration is based on sound ecological principles. Specifically he argues that by making the cows sacred, the Hindus conserve an ecological balance and population density that would otherwise be impossible to maintain. He is able to adduce some limited energetic evidence to support his argument. There is no doubt that Harris considerably altered our way of thinking about sacred cows; his formulation has forced researchers to take this phenomenon more seriously than they had done before. This alone is sufficient recommendation for Harris' argument, but my interest here is in the larger logic of his position and his general methodology as well.

After the sacred cows are disposed of, Harris moves on to other cultural practices for which he provides a cultural-materialist explanation. In comparing places where pigs are venerated with places where they are abominated, Harris rejects the "cultural" interpretations of these phenomena out of hand. Once again he equates the "natural" with the mundane and material and implies that culture is the opposite: unnatural, vague, spiritual (real nature versus unreal culture). He does sympathize a bit with Maimonides' medical materialism because it deals with the natural, definite, mundane forces that are involved in everyday life (p. 40). But Maimonides was too narrow for Harris; he did not include the whole ecosystem of which human behaviors in relation to pigs are a small part.

Given the desert environment of the Near East, Harris argues that pig prohibitions are good ecological practice. But apparently good ecological strategy alone is not enough for "pre-scientific" people, for "as in the case of the beef-eating taboo, the greater the temptation [here to raise pigs], the greater the need for divine interdiction" (p. 44). Thus people have to create elaborate false explanations to support their correct ecological strategies. This is the only way to avoid the temptation of destroying the ecological basis of their societies.

Just why people are tempted to do something wrong that the gods must protect against is not clear. If evolution is the guiding force in human behavior, why do people not simply behave as they must without any cultural "dreamwork"? Harris' answer apparently is that, in the absence of modern science, these ecological forces cannot be understood directly. Hidden in his apparent respect for the practices of other cultures is a unilineal argument for the development of objectivity and science, a view that has been immensely popular in the Western world for a long time. The Western world is placed at the pinnacle of rationality.

But if uniform material causes have uniform effects (as science insists), we are within our rights to ask what are the practical, natural, mundane forces that obligate Harris to seek this form of objectivity and that cause him to wrap social reality in science. His answer is that science is objective and that he is a scientist. So despite the fanfare, his argument ends up restating, though with some new twists, the contrast between the cultural orientations of primitive societies and modern ones. He believes that the full potential of modern society has not yet been realized and that its realization requires the intercession of scientific intellectuals. This is hardly a new idea.

When Harris deals with primitive warfare, his lack of regard for people's own conceptualizations of their lives and motives becomes even clearer.

> Irrational and inscrutable motives predominate in current explanations of primitive warfare. Since war has deadly consequences for its participants, it seems presumptuous to doubt that the combatants know why they are fighting. . . . But . . . the answers to our riddles do not lie within the participants' consciousness. The belligerents themselves seldom grasp the systemic causes and consequences of their battles . . . [P. 62]

People have motives for what they do but they are incapable of understanding the real reasons for their actions. These reasons are systemic, not individual or motivational. Only true scientists can comprehend systemic causes and consequences.

This position is rather awkward. Harris argues that people must have a structure of culturally created motivations that encourage them to behave as they must for the sake of their system, but their set of motivations never represents the system as it really is. To see the "real" system, an outside scientist is required. But by what evolutionary path can people end up incapable of understanding the "real" causes and consequences of their behavior? Surely people in other cultures are not less intelligent than we are.

Ecosystem Analysis

Harris claims that the key to all these problems is ecosystem theory, especially energy-flow analysis. He wants to show that all such behaviors are part of ecological adaptive strategies that yield the best possible results in their context. By applying these "material" principles to the analysis of unusual human behaviors, Harris tries to solve cultural riddles.

Does ecosystem analysis support him? The kind of ecological theory and the ecological data he uses are exceptionally primitive. The sacred cow argument is perhaps the one for which he has the best empirical evidence.

The portraits of the Indian ecosystem (pp. 16–19, 22) and of the others Harris discusses do not approximate any ecological standard. They are evocations of whole environments, incredibly diverse areas with enormous diversity of interlinked ecosystems. While there is no question that India, even the planet Earth, can be called an ecosystem, there is a real question whether this level of abstraction is relevant to a behavioral analysis such as the one Harris tries to make. Harris' argument centers on fine adjustment mechanisms in local ecosystems in which cows play a particular role. Yet as his portrait of the ecosystem is virtually India-wide, it is necessarily vague and abstract.

At the level of local detail, even ecologists who deal with relatively simple ecosystems find it necessary to devote hundreds of pages to the microenvironmental diversity and complex dynamics of an ecosystem before a small sample of the activities of humans can even be factored in (Netting 1971, E. Smith 1980, Winterhalder 1977). Thus Harris does not really apply ecological analysis to the Indian environment; rather he evokes the material world and then goes on.

Unfortunately this is true of all of Harris' cultural-materialist works, though as a matter of theoretical conviction he argues persuasively for the study of the specificities of the material interactions between humans and our environments. As he offers neither careful operationalization of concepts nor empirical proof, however, his argument can appeal only to those who are already convinced. While his audience is large and enthusiastic, the fact remains that Harris uses an advocacy method of argument when he deals with the human condition. This is not a scientific strategy.

The Sources of Scientific Objectivity

The question of objectivity is particularly vexing. In the analysis of sacred cows, for example, Harris squares off against Alan Heston, who has argued that the cows perform important ecological functions but that India would be better off with fewer of them. Harris counters that Heston's program would lead to the elimination of small farmers and the improvement of the lot of the larger farmers. Harris is certainly right to consider the distributive effects of development policies, but the difference between the two scholars opens up an interesting question. If Harris' cultural materialism is supposed to explain the sources of cultural ideas, then where do Heston's false ideas come from? And what is the source of Harris' correct ones?

Though Harris provides no clear answer in abstract terms, his explanation can be derived from the overall structure of his analysis. The Indian farmers' ideas arise from the cultural "dreamwork" that wraps and hides the "naked truths of social life." The penalty for thinking other than the way they do would be mass starvation. This much is consistent with Harris' cultural materialism.

In capitalist societies, and especially in the ranks of academe, people are capable to thinking up ideas and rules for behavior that are completely at odds with the "real" world. How they can do so in a world that, according to Harris, is uniformly ruled by material causes is a real problem for his theory. His answer appears to be that a short-lived bubble of capitalist abundance has somehow unhooked us from the real world. This is a major part of the argument in *Cannibals and Kings*. It appears that uniform material causalities argued for in Harris' cultural materialism are not so uniform after all.

Riddles that Dissolve into Social History

Harris' argument contains a source of slippage that effectively demolishes most of his own theoretical claims. At one point he says: "This is an appropriate moment to deny the claim that all religiously sanctioned food practices have ecological explanations. Taboos also have social functions" (p. 45). Though not particularly debilitating in this context, this point comes up repeatedly in his work. While Harris insists on the universal applicability of his materialist arguments, whenever he runs into trouble he invokes a social or historical factor to explain the anomaly. This strategy effectively insulates his theory against any negative evidence, much as does Wilson's use of proximate and ultimate causality. So much for science.

In dealing with primitive peoples Harris introduces historical forms of explanation in a most disturbing way. Attempting to explain anomalies in the behavior of the Yanomamo Indians of Venezuela, Harris argues that their recent movement into their geographical area and adoption of a new subsistence system accounts for many of their problems. He speculates that they had been a nomadic people and began only recently to experiment with agriculture. The resulting great food increase led to higher population densities, which in turn created problems with hunting and other activities. The Yanomamo "have already degraded the carrying capacity for their habitat" (p. 105).

But how are such processes possible in the framework of his theory? How can the "adaptedness" of one system be compared with another? How can the degree to which a particular society meets a set of analytical expectations or deviates from them be measured? Without having answered these questions, Harris deviates from direct cost-benefit optimization in the analysis of particular behaviors or groups of them whenever it suits his convenience to do so.

This is not a side issue. If we follow Harris' approach to the analysis of a particular society, we must know whether or not the society has been in its present location long enough to have worked out a well-balanced adaptation to the environment. Such knowledge requires some useful measure of "adaptedness." Neither of these requirements is met. Any time the data do not fit his expectations well, the lack of such knowledge constitutes an open invitation to consider that the cause is historical, that the deviation is caused by some interference. But the deviation could also be caused by bad data, poor formulation of the analytical categories, or simply an incorrect analysis of the data.

An evolutionary ecologist, recognizing that these issues have to be fully settled by agreed-upon measurements, would see this requirement as a heavy additional weight on an analytical framework that is already difficult to apply because it makes extraordinarily comprehensive empirical demands. Harris does not face up to this problem. His approach is to tell just-so stories about adaptation without much interference from the data. Lacking good ecological data, the formulation of alternative hypotheses, consistent analytical standards, and operational tests, Harris' view becomes a textbook case of adaptationism.

Examples of this approach are found throughout the book. Harris argues that the potlatch of the Kwakiutl functioned as a necessary redistribution of resources; that is, it was a systematic process. Why did the Kwakiutl have such a system when the Yanomamo do not? Harris would probably answer that the Kwakiutl had been in their environment longer and had worked out their adaptation fully. But what theory of evolutionary ecology tells us how long it takes to become "adapted"?

While it makes sense to argue that societies that have taken up a new way of dealing with the environment relatively recently will be less finely tuned to the local ecology than those that have been in place longer, such an argument must meet high theoretical standards. Among other things, there must be a set of definite criteria by which to judge the "adaptedness" of societies. Further, it is necessary to model the adaptedness of a society over time and predict how long is long enough for stable adaptations to develop in particular ecological zones given certain sets of food-getting strategies.

This line of reasoning also implies that some kinds of societies degrade their environments and others do not. The theory must explain under what circumstances a society does and does not produce a stable adaptation. Harris opens the door to all of these dilemmas the moment he abandons direct material causation by invoking a loose historical standard for judging adaptations.

The alternative is even worse. To argue, despite the existing evidence, that all societies are well adapted to their ecosystems would be a travesty. But to save the argument by shunting the deviations from expected adaptations off to history is no solution. Harris' next book, *Cannibals and Kings*, attempts to solve just this problem.

As the reader is brought to consider messiahs and the witch panic in Europe, the problems of historical causality get worse. Harris uses these cases to try to explain how consciousness got so far out of touch with "reality" in Western societies. His strategy obligates him to explain why the panic broke out when it did and not earlier or later. This sort of question plagues all historians and is not helped a bit by Harris' cultural materialism.

Harris seeks demographic and ecological causes for these events, but his arguments rest on such a long series of assumed relationships that they are of little interest to anyone familiar with the details of the great religious upheavals in Europe and the United States. If Harris only wanted to persuade us to pay close attention to the material aspects of social life during the period of these outbreaks, no one could disagree. But he sacrifices the analysis of the detailed social etiology of these movements in order to make room for his particular materialist interpretation, which then turns out to be too vague to be helpful.

"The Return of the Witch"

As he begins to close out the argument, he speaks metaphorically of the "return of the witch," meaning the ways in which contemporary consciousness is out of touch with "practical and mundane constraints" (p. 252). How can consciousness, which is, in Harris' own words, "adapted to the practical and mundane conditions" (p. 253), get out of touch with those conditions? No answer is given.

The reader is left with the idea that primitives think about *what is* in fantastic symbols while peasants and academics think about *what is not* in other kinds of fantastic symbols. Only scientists, and in particular cultural materialists, relate "true" consciousness to the practical and mundane constraints of everyday life. Marvin Harris believes that we must use science to learn to eliminate from our culture all of our false conceptions of how things are and ought to be.

Harris' drive for data about the material world is good, a useful corrective to anthropology's large dose of idealist bias and empirical laziness. His inclusion of the Western world in anthropological comparative statements is important and his flashes of insight and wit are engaging. Still the book is a great disappointment. Ecology is invoked only in the loose sense that everything must have energy costs and benefits. Harris engages in an extreme form of adaptationism, creating adaptive stories unmediated by any sense of the operational requirements of ecology. This approach hardly represents the sophistication of evolutionary ecology and the multiplicity of ways its theories and methods could be adapted to anthropological use.

The book does not cope with the central theoretical problem in attempts to deal with humans ecologically: how to treat the interaction of economic and energy currencies, since they do not match up in any obvious ways (Bennett 1976, E. Smith 1980). Because of culture, human ecosystems are not bounded in space and time in the same way that nonhuman systems are. The kinds and amounts of information they process are different. No account is taken of this fact.

Instead Harris tries to make history the arbiter of adaptation, arguing that a certain amount of time is required for a system either to settle adaptively into a stable strategy or to disappear. No means of determining the amount of time needed is ever suggested, or any means of distinguishing between systems that are maladaptive and those that are not yet adapted. These questions require answers from any theory that calls itself scientific.

Harris' view of human history is not nearly so new or revolutionary as he thinks it is. He treats tribal societies as ones where people do what they must in a balanced relationship with the environment, avoiding the temptation to do otherwise by wrapping their actions in fantasy. In early states, people behave pretty much as in tribal societies, but the logic of political power forces a certain degree of environmental modification. In capitalist states, people are driven to deplete the environment by the logic of capitalism, while their consciousness of the situation is the exact opposite of what is really happening. The ideal future society is one in which scientific consciousness of the ecological and economic constraints at the base of all societies will predominate. Through this knowledge a reconciliation of consciousness and reality will be created so that the dichotomy between nature and culture can be abolished. This is a rather commonplace form of utopian rationalism.

Cows, Pigs, Wars, and Witches leaves the following problems unresolved: What causes the major transitions in human history? What is the relationship between consciousness and ecology? What is adaptation? How can we distinguish between systems that are maladaptive and systems that simply have not yet achieved an adaptive balance? What is the source of scientific objectivity? *Cannibals and Kings* and *Cultural Materialism* are Harris' attempts to address some of these questions.

Cannibals and Kings

Cannibals and Kings, subtitled *The Origins of Cultures*, shows that Harris is aware of the major flaw in *Cows, Pigs, Wars, and Witches*:

the lack of a dynamic that moves history through the transitions he has described. This work is intended to provide that dynamic. He states the argument succinctly at the outset:

> In the past, irresistible reproductive pressures arising from the lack of safe and effective means of contraception led recurrently to the intensification of production. Such intensification has always led to environmental depletion, which in general results in new systems of production. . . . [Harris 1977:xi]

Then he brings his policy position forward:

> That a blind form of determinism has ruled the past does not mean that it must rule the future. . . . I have no difficulty in believing both that history is determined and that human beings have the capacity to exercise moral choice and free will. . . . In my opinion, free will and moral choice have had virtually no significant effect upon the directions taken thus far by evolving systems of social life. . . . It behooves those who are concerned about protecting human dignity from the threat of mechanical determinism to join me in pondering the question: why has social life up to now consisted overwhelmingly of predictable rather than unpredictable arrangements? I am convinced that one of the greatest existing obstacles to the exercise of free choice on behalf of achieving the improbable goals of peace, equality, and affluence is the failure to recognize the material evolutionary processes that account for the prevalence of wars, inequality, and poverty. [pp. xi–xii]

Cannibals and Kings provides a population/environment motor to drive the cultural evolutionary process along. The book also clarifies Harris' ethical stance considerably. Equating rationality, knowledge, and the exercise of freedom, he seeks to study the laws of nature in order to defeat them or at least to subordinate them to certain ethical standards.

The Cultural Basis of Cultural Materialism

Harris' population/environment argument, strangely enough, rests on a cultural foundation, though apparently he does not see it as such. He claims that the severe costs of infanticide to humans are the real motor of human history. It is very difficult for humans to endanger pregnant women's lives and to kill children. But he does not explain why the killing of infants is worse for humans than for other animals. Nor does he argue that it is harder on human females physically. Harris believes that infanticide has a high moral and cultural cost.

For this to be the case, Harris has to argue for certain panhuman cultural capacities that have not evolved over time. The high cost of infanticide is treated as a universal, fixed characteristic of the human species as a whole. This generic statement about what is "natural" to humans is undefended and is incompatible with Harris' evocation of an evolutionary view. In such a view, a fixed, universal "human nature" has no place.

"I suspect," Harris writes, "that only a group under severe economic and demographic stress would resort to abortion as its principal method of population regulation. . . . In the case of both geronticide and infanticide, outright conscious killing is probably the exception" (p. 15). His concept of "costs" of population control is clearly a culturally mediated one, resting on a view of panhuman moral sentiments, an immensely popular idea. Harris is an optimist about human nature, a point relevant to much of his popularity because audiences respond favorably to it.

Adaptationism

Cannibals and Kings is even looser than the previous work in its appeal to adaptationist stories. Harris states, "Yet I have already *shown* that what keeps hunter-collectors from switching over to agriculture is not ideas but cost/benefits. . . . This theory *explains* why the domestication of plants and animals occurred at the same times and places in the Old World" (pp. 26–27; emphases mine). "I have shown" would be more accurately stated as "I have argued." As for the claim that the theory "explains" the timing of domestication, Harris offers a hypothesis, then takes it as confirmed, and then converts it into a theory that explains. This indefensible strategy is consistently pursued throughout the book.

In discussing the origin of war, Harris argues that variations in the intensity of war are caused by cultural factors. "Obviously it is part of *human nature* to be able to become aggressive and to wage war. But how and when we become aggressive is controlled by our cultures rather than by our genes" (p. 37; emphasis mine). This kind of argument posits a timeless human nature apart from history. It also embodies an extreme form of environmentalism: humans have the capacity to do many things but the particular environment *absolutely* determines what they will do. Soon Harris takes the next step: "Warfare ... is not the expression of human nature, but a response to reproductive and ecological pressures. Therefore, male supremacy is no more natural than warfare" (p. 57).

Human nature is constant but its expression varies in accordance with the situation. If human nature is constant, then how can Harris have any hope that humanity will change in the future? He must call on science to create an environment that will permit the expression of the "true human nature" (as opposed to the nature of observed humanity). This position is neither new nor defensible according to the cultural-materialist ground rules he lays down. It is familiar to us from Hippocrates and Jean Bodin, among others.

Material Causes, the Human Will, and the Ethical Duty of Science

In regard to causal statements, Harris' lack of attention to operational questions places him in an awkward position. He states, for example, "The Oedipus complex was not the cause of war; war was the cause of the Oedipus complex (keeping in mind that war itself was not a first cause but a derivative of the attempt to control ecological and reproductive pressures" (pp. 65–66). Then he justifies this statement with the following: "It is an established principle in the philosophy of science that if one must choose between two theories the theory that explains more variables with the least number of independent unexplained assumptions deserves priority" (p. 66).

This, of course, is true, but he has left out a step. Ockham's razor can be invoked only in operationalized explanations. Since statements such as those about the Oedipus complex are not in any operational form, it is impossible to decide which alternative view does contain the largest number of unexplained assumptions. Harris rarely moves to the level of operational research; yet the issues he tries to resolve cannot be dealt with by logical manipulation alone.

Other statements move us toward an even more ambiguous stance on causality:

> War and sexism will cease to be practiced when their productive, re-productive, and ecological functions are fulfilled by less costly alterna-tives. Such alternatives now lie within our grasp for the first time in history. If we fail to make use of them, it will be the fault not of our natures but of our intelligence and will. [P. 66]

Here the separation of will and intelligence from human nature must be kept alive if the dynamic of his model of history is to work. Again we are required to view human nature as outside of history—an en-terprise that makes no biological sense. Yet without such a view Har-ris' moral claims lose much of their support, as in the following case:

> I urge those who feel that my explanation of the evolution of culture is too deterministic and too mechanical to consider the possibility that at this very moment we are again passing by slow degrees through a series of "natural, beneficial, and only slightly . . . extra-legal" changes which will transform social life in ways that few alive today would con-sciously wish to inflict upon future generations. Clearly the remedy for that situation cannot lie in the denial of a deterministic component in social processes; rather, it must lie in bringing that component into the arena of popular comprehension. [P. 82]

By bringing these determinisms to the attention of the people, he hopes to improve the situation. Human nature is assumed to be good and reasonable; humans, faced with the right information, are likely to make much more constructive decisions than they have done in the past. As I said, Harris is an optimist.

And then he hedges: "I do not claim that the analysis of ecological costs and benefits can lead to the explanation of every belief and prac-tice of every culture that has ever existed" (p. 137). In the absence of operational definitions, this kind of caveat becomes the ultimate fudge factor. It says that the explanation explains what it explains and does not explain what it does not explain.

Finally, Harris derives a lesson from Karl Wittfogel:

> The effective moment for conscious choice may exist only during the transition from one mode of production to another. After a society has made its commitment to a particular technological ecological strategy for solving the problem of declining efficiency, it may not be possible to do anything about the consequences of an unintelligent choice for a long time to come. [P. 163]

This is quite an important point, for it adds to his earlier model of societies that are adapted and others that are not the idea that there are certain open doors in history. Only during these transitions is free will operative. This is Harris' way of reconciling his cultural materialism with his appeal to free will. It is not clear that the reconciliation makes any sense.

Harris believes that we are at such an open juncture now; thus the exercise of informed free will is crucial at present. He states:

> No one who detests the practice of kowtowing and groveling, who values the pursuit of scientific knowledge of culture and society, who values the right to study, discuss, debate, and criticize, or who believes that society is greater than the state can afford to mistake the rise of European and American democracies as the normal product of a march toward freedom. [P. 175]

How, then, do we keep them from disappearing?

> Only by decentralizing our basic mode of energy production . . . can we restore the ecological and cultural configuration that led to the emergence of political democracy in Europe. This raises the question of how we can consciously select improbable alternatives to probable evolutionary trends. . . . To change the world in a conscious way one must first have a conscious understanding of what the world is like. . . . It is only through an awareness of the determined nature of the past that we can hope to make the future less dependent on unconscious and impersonal forces. . . . While the course of cultural evolution is never free of systemic influence, some moments are probably more "open" than others. The most open moments, it appears to me, are those at which a mode of production reaches its limits of growth and a new mode of production must soon be adopted. We are rapidly mov-

ing toward such an opening. . . . In life, as in any game whose outcome depends on both luck and skill, the rational response to bad odds is to try harder. [Pp. 194–96]

This idea of the open moment and the appeal to democratic values together provide the drama and the call to action that makes Harris a compelling writer. But it does not make him a materialist or an evolutionist in any clear sense. Indeed, we have heard similar arguments before, in the writings of Jean Bodin and a host of other preevolutionary social reformers.

Cows, Pigs, Wars, and Witches Revisited

Does *Cannibals and Kings* correct the weaknesses of *Cows, Pigs, Wars, and Witches*? It does not. The use of adaptationist arguments is more rampant and attention to operationalism is nil. Even the occasional encouragements to readers to get the data needed to examine these general propositions have pretty well disappeared.

The discussion of the relationship between culture and biology is not moved forward from the position taken in the earlier book, though it is restated in clearer terms in *Cannibals and Kings*. Human nature is given a definite static quality that it did not have earlier, a quality that eliminates much of the possibility of thinking about humans in evolutionary terms. Harris plainly creates a definitional separation between human nature (which is static), human cultures (which evolve), and human will and intelligence (which, though part of human nature, may or may not be exercised according to principles that Harris never clarifies).

The stages of human history are partly explained by means of Harris' abstract model of population increase, intensification, and depletion. This abstract model is quite interesting. With attention to the variety of operational problems involved and the elaboration of serious research hypotheses, this kind of model could be employed and its usefulness could be assessed. But Harris does not rely on the model as much as he claims. The rampant adaptationism of his mode of argument prevents deployment of the model or even discussion of the problems of deployment. He attempts to settle the matter in favor of the model with appeals to inadequate data from the research of others.

Though at first glance the model appears to be a materialistic one, it relies on the permanence of the values humans place on human life, especially on the lives of mothers and children. Were it not for these values, nothing would prevent abortion and infanticide from solving the population problem. Thus Harris' argument ultimately rests on assumptions about panhuman moral preferences. While these assumptions may be correct, they must be argued forthrightly as part of his model. In actuality Harris has created an eclectic model of human behavior and history, despite the aspersions he casts upon eclecticism.

Finally, the sources of scientific objectivity, a problem in the first book as well, are not clarified at all. To this problem is added the confusion surrounding the desirability of democracy and rationality as guiding principles of human life. Rather than take a more clearly moral position regarding both democracy and rationality, Harris suggests that democracy arises under particular ecological conditions and that these conditions have to be reproduced if democracy is to be preserved. And he argues not that rationality should be an ethical standard but that rationality is scientific and science gives us control over our environment. Thus rationality can be justified because it is evolutionarily successful.

In both of these arguments Harris moves from "is" to "ought" without being aware that he is doing so. He continues to claim that evolutionary biology can guide us into a rational assessment of our situation, and that once we have made that assessment, we will know exactly how to behave. And if we fail, he says, it will be a failure of the will and intelligence. Thus we shall be to blame for not following our material interests.

Cultural Materialism

To build a "scientific" basis for this view, Harris then wrote *Cultural Materialism* (1979). This book attempts an epistemological justification of the cultural-materialist strategy, both in its own right and in comparison with all other major strategies of culture analysis. Though Harris continues to write on the subject, the trajectory leading from *The Nature of Cultural Things* (1964) to *Cultural Material-*

ism forms a remarkably complete corpus containing methodology, history of theory, synchronic and diachronic theory, and epistemological justification. Indeed, the overall coherence of his enterprise is intriguing.

Cultural Materialism is not very successful in accomplishing its aims because it does not effectively address most of the issues the earlier books leave unresolved. Instead it sets out a conventional philosophy of science and then proceeds to heap scorn on non-cultural-materialist approaches to the study of human behavior. Though often entertaining—few writers can match Harris' way with words—these critiques do not advance the cause of materialist or evolutionary analysis in any clear way.

Cultural Materialism as Science

Harris makes quite acceptable general statements about science. He stresses the openminded comparison of alternative theories and argues that a scientific strategy should be explicit regarding the epistemological character of its basic variables, the relationships between the variables, and the interconnected bodies of theory that are relevant to it. He also stresses parsimony in theory formulation and the continual monitoring of theory through empirical testing. No one could disagree that such an enterprise is laudable and perhaps possible. It does not describe what Harris has done in his previous works.

The sections on cultural materialism proper add nothing to the doctrine that has not already been heard before. Harris stresses the distinction between emic and etic data and the importance of measuring the discrepancy between them. And he argues that "the universal structure of sociocultural systems posited by cultural materialism rests on the biological and psychological constants of human nature, and on the distinction between thought and behavior and emics and etics" (Harris 1979:51).

Harris endeavors to incorporate the concept of infrastructure into his argument both to clarify his position and to incorporate those elements of Marxism that he deems useful. Infrastructure, he says, is the interface between nature and culture, and then he states: "Unlike ideas, patterns for production and reproduction cannot be made to appear and disappear by a mere act of the will" (p. 58). Where now is the power of the will (and intelligence) claimed in *Cannibals and Kings*?

Individual Wills and Historical Trajectories

When Harris turns to the problem of why the sum of individual biopsychological utilities, calculated on their own, will not yield a predictive theory of cultural evolution, his explanation sounds like a negative model of group selection in which groups are much less rational than individuals.

> The more hierarchical the society with respect to sex, age, class, caste, and ethnic criteria, the greater the degree of exploitation of one group by another and the less likely it is that the trajectory of sociocultural evolution can be calculated from the average bio-psychological utility of traits. This leads to many puzzling situations in which it appears that large sectors of a society are acting in ways that diminish their practical well-being instead of enhancing it. [Pp. 61–62]

Decaying Infrastructures and Cultural Mystifications

This idea is quite important for his model because it is the beginning of an attempt to take account of a problem I stressed earlier. He is trying to explain why tribal societies' adjustments to their circumstances break down as social stratification develops.

Later he claims that decaying infrastructures yield the worst forms of ideological mystifications. "A final ideological product of a decaying infrastructure . . .[is] the growing commitment of the social sciences to research strategies whose function it is to mystify sociocultural phenomena by directing attention away from the etic behavioral infrastructural causes" (p. 113).

Harris has finally diagnosed the cause of the "overdose of intellect" he decried in *Cows, Pigs, Wars, and Witches*: it is the collapsing infrastructure of capitalist industrial society. This decay is causing our consciousness to stray from the real problems we face. Without expert guidance from intellectuals who study the infrastructure directly, we will not survive. This is why we think about what is not rather than about what is.

Cultural materialism is posed as the answer to this dilemma. How we know when an infrastructure is decaying is not addressed.

Eclecticism and Obscurantism

Among all analytical approaches, Harris most dislikes what he calls eclecticism. He is particularly vexing on this point since he does not distinguish between eclecticism and confusion. He begins by claiming to have discovered that eclecticism is itself a strategy of analysis and then, by fiat, he says that "eclecticism cannot lead to the production of a corpus of theories satisfying the criteria of parsimony and coherence" (p. 288).

This statement reveals much about Harris. Such a bald assertion can be based only on a metaphysical belief that the world of observation operates according to a few simple, regular laws. This view cannot be derived from the principle of parsimony because that principle calls not for the simplest explanation but for the simplest *possible* explanation. If there is reason to think that the empirical world operates with a set of heterogeneous causes that may not be usefully reduced to each other—biological and cultural causes for example— then a parsimonious and coherent explanation would have to be "eclectic" in Harris' terms.

He argues that eclecticism would be viable only if nature were fragmented and inconsistent (p. 290), makes undefended metaphysical distinctions between crucial and less important variables (p. 295), and finally equates eclecticism with confusion:

> The notion that all the parts of sociocultural systems are equally determinative of each other is a prescription for theoretical chaos. . . .There is as little room in the social sciences for the idea that all parts of socio-cultural organisms "inneract" [sic] equally, as there is room in physiology for the belief that all parts of a plant or animal are equally vital for the maintenance of life functions. [P. 312]

This is a red herring. An eclectic argument in no way must assert that all parts of a system are equally determinative; rather eclectic explanations emphasize that different causes are determinative in different degrees under specifiable conditions. Since Harris' own model includes demographic variables, panhuman nature and ethics, and intelligence and will, it seems to me that he is thinking eclectically himself. Harris does himself a disservice by deemphasizing precisely this eclectic component of his own vision.

Harris' criticism of Marshall Sahlins' rejoinder to his analysis of Aztec cannibalism (Sahlins 1978) shows just how pointless this kind of debate can be. According to Harris, Sahlins "has no alternative explanation. The sole purpose of his unremittingly negative critique is to prove that Aztec 'culture is meaningful in its own right,' a proposition to which one cannot object but which has no bearing on the question of whether or not Aztec cannibalism can be explained by cultural materialist theories" (p. 339).

Harris objects to Sahlins' departure from cultural materialism—by which Harris here means demographic/ecological causalities from which other cultural phenomena are derived. Yet Harris' cultural materialism contains a variety of heterogeneous and untestable assumptions about panhuman ethics and morality which drive his whole model. There is nothing less empirically testable or more eclectic in Sahlins' assertion about culture than in Harris' assertions about human nature and values. If cultural systems are meaningful in their own right, and we agree that such systems arise historically and maintain a certain coherence over time, then we can study them systematically and historically in conjunction with the infrastructure without creating any contradictions. The issue between Harris and Sahlins is not science but metaphysics.

America Now

Harris' more recent book, *America Now: The Anthropology of a Changing Culture* (1981), claims to extend his techniques directly to the study of American society. It begins by invoking the collapse of the American dream: "This is a book about cults, crime, shoddy goods, and the shrinking dollar. It's about porno parlors, and sex shops, and men kissing in the streets. It's about daughters shacking up, women on the rampage, marriages postponed, divorces on the rise, and no one having kids . . ." (p. 7). Harris is characteristically interesting and entertaining, offering some insightful observations about American life. But there is no connection between his analysis and any applications of energy-flow analysis. The political and moral values that motivate Harris are more clearly in evidence here than ever before.

Harris claims that "traditional moral and spiritual values have lost their appeal"; it is the function of the book to explain why. According to Harris, it is best to start analyzing such problems from the bottom up, "from the changes in the way people conduct the practical and mundane affairs of their everyday lives" (p. 11). But rather than relying on a strong form of techno-environmental determinism, Harris hedges by stating that "there is no single chain of causes and effects that can be followed out link by link from one basic change to all others" (p. 12). He sets out to show instead that the whole array of changes fits together in an intelligible pattern; other thinkers, he says, see these problems as unconnected or as the workings of obscure forces. "The task of this book," he writes, "is to reassert the primacy of rational endeavor and objective knowledge in the struggle to save and renew the American dream" (p. 15). The use of rationality to revitalize democracy, a theme lurking in most of his other works, now takes pride of place.

Throughout he takes up issues that all social commentators on the American scene have examined: shoddy goods, poor service, economic problems, the women's movement, gay liberation, crime in the streets, new religious cults. The subjects are interesting and Harris is good at picking out striking details. Yet even the best of the chapters reads like an analysis in the editorial pages of the *New York Times* rather than an application of a "science of culture." Nothing in this supposed application of cultural materialism seems to produce insights different from those to be found in standard liberal, conservative, and Marxist critiques. Certainly the connection between this analysis and ecosystems analysis has been severed; in its place stands an invocation of mundane conditions and general patterns of change.

In the end Harris' politics and moral aims dominate all other motives:

Given the enormous power and formidable inertia of the hyper-industrial oligopolies and bureaucracies, there is only a slim chance of achieving a future more in accord with the vision of freedom and affluence on which past generations of Americans were nourished. Nonetheless, this chance is sufficient to support a rational hope of reversing the trends that have led to America's present malaise. The will to resist and to try for something better is an important component in the struggle against oligopoly and bureaucracy. Of course, to desire something strongly enough to fight for it does not guarantee success. But it changes the odds. The renewal of the American dream may be improbable, but it will become finally impossible only when the last dreamer gives up trying to make it come true. [P. 183]

The preservation of America (not unlike the preservation of our species for Wilson) through rationality is the goal. Harris clearly claims that dreamers can affect events, a view that lurks in all of his works under the mantle of ecologism.

Harris equates utopia with a state in which rationality is used in the service of democracy. It is a society in which everything works, people are supportive of one another, the economy is in balance, family structure is stable, the crime rate is low, and irrational religious sects are on the wane. Somehow his long intellectual detour through evolutionary biology and the science of culture has ended up reproducing a conventional middle-class version of American life as the ideal.

Conclusions

Major metaphysical assumptions are necessary for the operation of Harris' model. He asserts that reality is orderly, causally uniform, and divided directly into the following dichotomous properties:

Nature	Culture
law	will
etic	emic
genetic	cultural
natural	artificial
mundane	transcendent

For Harris, human nature is constant and uniform over time and space. Humans have the capacity to discover directly the order that causes reality to operate as it does. Equating science with objectivity about reality, he argues that science demands that cultural systems be derived logically from natural systems.

Harris divides evolution into biological and cultural evolution, yet nothing in his theory provides a basis for such a distinction. He unproblematically considers biological evolution to be an optimizing process and treats selection as a constant force. His implicit ethical stance is democratic, with a sub rosa requirement that democracies be guided by scientist/kings—presumably well versed in cultural materialism.

Finally, he argues that natural causes account for culture as a general human characteristic and for the details of cultural systems. Human history begins in a Malthusian balance but larger stratified societies develop internal contradictions that can be dealt with only through policy. Our failure to understand this predicament is caused by our decaying capitalist infrastructure.

Since Harris' claim to science is based on his evocation of evolutionary biology, and specifically of energy and demographic analysis taken from ecology, it is legitimate to ask how well he has represented these theories in his work. The answer is not well at all. The metaphysical assumptions of evolutionary biology do insist that nature is orderly and subject to the constant action of uniform causes, but there is no evolutionary biological distinction between the material and spiritual aspects of human behavior. This is Harris' addition.

Evolutionary biologists who treat these issues carefully would claim that while it must ultimately be possible to reduce culture to nature, reduction and explanation are not the same thing (Hull 1974). The reduction of culture to nature cannot explain the operation of cultural systems—it only sets broad parameters within which culture exists. Further, most evolutionary biologists would certainly agree that there is no acceptable experimental evidence that could lead to the formulation of any general propositions about "human nature."

There is no basis in evolutionary theory for separating biological and cultural evolution. The concept of cultural evolution is a misleading analogy based on a misunderstanding of biological evolution

(Greenwood and Stini 1977). There is one evolutionary process—biological evolution, of which culture is a part. Nor is evolution an optimizing process (Gould and Lewontin 1979). Selection is neither constant nor unitary; it is episodic and focused on certain traits and certain moments (Gould and Eldredge 1977). Finally, most evolutionists feel that evolution does not provide any clear source of ethics. Little of Harris' theory is directly implied in evolutionary biology; most of what he says runs counter to the core of biological science. He did not need evolutionary biology at all for the formulation of his cultural materialism.

The source of Harris' theory is found in pre-evolutionary thought, in which a radical dichotomy between nature and culture was held to be scientifically meaningful and politically useful. The relationship between nature and culture was seen as one of struggle between natural laws and human will, and theorists used naturalistic arguments to set ethical and political standards. Harris' appeal to rationality, will, and intelligence belongs to this tradition, not to the tradition of Darwin.

To understand Harris' views, a knowledge of the works of Hippocrates, Bodin, and Torres Villarroel is more helpful than a reading of Darwin. The conflict between the genealogical and environmental principles and the political management of states is a key theme for Harris. With Torres he shares the basic underlying notion that a series of fundamental and stable moral premises undergird "human nature" and that our relationship with nature can either support or destroy these premises. Torres' desire to have people recognize their basic constitutions and to harmonize themselves with "nature" is virtually identical to Harris' plea for a rational approach to the problems of population, pollution, and war. Torres' scathing critique of abstract intellectualism is closely echoed in Harris' commentaries on contemporary social scientists, humanists, and politicians. All these thinkers find sermons in nature in ways that should make evolutionary biologists shudder.

CONCLUSION

The Unmet Challenges of Evolutionary Biology

It is clear that a surprising number of pre-evolutionary elements persist in contemporary works on "human nature." But to highlight these continuities is not to explain them; explanation is a much greater challenge.

Partly their persistence is a result of the immense staying power of major cultural systems. The forces of cultural continuity and the power of ideas are much greater than many contemporary scientists realize. But ideas do not persist in a vacuum, and on occasion they do change abruptly. Persistence is to be explained, not merely invoked.

There are basically two reasons why this nonevolutionary view of nature persists, both stemming from the social uses made of the relationship between nature and culture. Politically the people who vie for social power virtually always attempt to root their claims for the rightness of their views in assertions about "nature." The argument generally goes that natural laws must be obeyed lest we destroy our relationship with nature and disappear.

The position of greatest political power in society is that of interpreting just what the laws of nature are and what responses they require from us. As theorists from Jean Bodin to Milton Friedman would have it, rulers must understand the laws of nature, which are beyond human control and volition. Nature cannot be changed. Only in the realm of culture (nonnature) do rulers have freedom of choice and thus also moral responsibility. Effective politics consists in knowing the difference between nature and culture and convincing the governed that natural laws, or previous rulers' ignorance of them, are to blame for the untoward characteristics of society.

Related to this general political matrix is the long-standing demand in Western society that all important political and moral obligations be based on natural laws. I am convinced that many who attempt to root their political and moral beliefs in nature do so in an effort to protect their beliefs either from direct attack or from erosion through social negotiation. By arguing that beliefs have a natural basis and using science to back them up, they try to place their own preferences beyond dispute.

The alternative is apparently too frightening to bear. If political and moral action is not based on natural laws, then society may be reduced to an arena of power struggles in which values not backed up by power cannot be expected to survive. This view has led some people to argue that, while they believe in evolution themselves, they do not think it should be taught in the schools. They feel that most people are not capable of leading moral lives without the support of superstitious beliefs about nature. This attitude is also the source of opposition to attempts to place a monetary value on such things as human life and the beauty of nature for the purposes of cost-benefit analysis; any such attempts, they believe, will automatically cause the perceived value of life and natural beauty to decline.

Behind all these debates lies our apparently unshakable belief in the political and moral importance of "natural laws." Why do we hold such a belief? Some scholars, including both Wilson and Harris, would argue that such beliefs have to be explained in material terms. I reject that proposition. Cultures differ. Just as languages separate and continue to diverge because of the differing assignments of meaning to sounds, so cultures diverge from one another. Once a culture has been formed, its patterns persist, giving its historical trajectory a kind of coherence. While cultures do not mechanically reproduce themselves over time, there is a strain toward consistency that often has amazing endurance, especially in critical areas of social concern. What matters is not the origin of the particular set of beliefs under consideration here but their effects on our perceptions and actions; the way social groups benefit from them; and the institutional structures they support or revolt against.

Ironically, it is precisely Western culture's emphasis on the political and moral importance of natural laws that has led us to devote so many of our resources to the study of nature. Our dedication to the sciences, physical and biological, derives from the belief that nature is the book in which political and moral directives are to be read. To study nature is to promote responsible citizenship.

Unfortunately, the accumulation of evidence about nature and the development of an understanding of the operations of the physical and biological world have led to the realization that the world does not tell moral tales. The Copernican revolution, the laws of thermodynamics, relativity theory, and evolutionism provide a picture of the world that does not give us political or moral instruction.

Though all of these scientific findings have had profound effects on our political and moral thought, evolutionary theory is potentially the most deranging development of all. Without fixed natural categories, without a fixed boundary between nature and culture, without a fixed "human nature," and without any overall direction in the life process, it is impossible to make nature into a source of ethical and political prescriptions. We have invested great intellectual, emotional, and financial resources in the study of evolution only to find the results unpalatable.

The challenges provided by evolutionism can be concisely stated. Evolutionism has challenged us to reconceptualize our views of nature and natural processes. It requires that we develop a view of human nature that does not radically divide biological and cultural forces. Finally, it forces us to find justification for our political preferences and moral beliefs somewhere other than in nature (and by implication, somewhere other than in science). Some of these challenges have been at least partially met; others have not.

Genuinely evolutionary views of the biological world have existed since Darwin's time and have continuously gained in sophistication as new discoveries in genetics, ecology, molecular biology, ethology, and paleontology have been incorporated into the evolutionary synthesis. One by one, the primary difficulties that faced evolutionary thinking have been overcome. The writings of such scholars as Charles Darwin, Thomas Huxley, G. G. Simpson, Theodosius Dobzhansky, Ernst Mayr, Stephen Jay Gould, and François Jacob are testimony to the power of the evolutionary view.

It may be argued that evolutionary thinking is now so complex and recondite that it can be appreciated only by experts. It is for this reason that pre-evolutionary thinking persists. This is not so. A perusal of François Jacob's brief book *The Possible and the Actual* (1982) shows how easy the evolutionary view is to understand (though not necessarily to accept). In the space of sixty-eight pages Jacob lays out an elegant synthesis of evolutionary thought that takes account of the problem of levels of organization of matter, the relationship between possible and actual organisms, and the essentially historical character of the evolutionary process. Thus clear syntheses of the evolutionary view are available. The problems discussed here arise not from the complexity of the evolutionary view but from an active desire to avoid some of its implications.

Another argument, one put forward by such scholars as E. O. Wilson, is that a genuinely biocultural view of humans has not yet been formulated, at least not in a way consistent with evolutionary theory. The lack of such a view is said to be the cause of our confusion. Despite its apparent plausibility, this claim is false.

Evolutionary views of humans encompassing the entire range of cultural activities have indeed been developed. James Spuhler's *Evolution of Man's Capacity for Culture* (1959) opened up a significant discourse on the evolutionary interactions between human biology and cultural behavior. This view has been expanded by Clifford Geertz (1973a, 1973b) and by many other anthropologists.

Gregory Bateson, in *Mind and Nature* (1979), uses the concept of patterned connections to trace relationships between the organic and inorganic worlds and the operations of the human mind. Though complex and unconventional, Bateson's formulation encounters no gaping impossibilities, no line between nature and culture that cannot be bridged. The same is true of Mary Midgley's *Beast and Man* (1978),

Vernon Reynolds' *Biology of Human Action* (1980), and William Durham's *Coevolution* (forthcoming).

If such biocultural perspectives exist, why do E. O. Wilson, Marvin Harris, Robert Ardrey, Konrad Lorenz, Desmond Morris, Richard Dawkins, and many others claim that they have discovered the only truly scientific evolutionary view of humans? The reasons are political and moral: the biocultural views of Spuhler, Geertz, Bateson, Midgley, Reynolds, and Durham do not provide moral prescriptions.

That a host of acceptably biocultural views of humans exist, none contradicting the basic tenets of evolutionary theory, has not been noticed by Wilson, Harris, and the others because they are not really interested in biocultural evolutionary views of humans. What they seek is to root their political and moral preferences in "science." If biocultural views do not provide answers to the problems of species survival, do not offer immutable categories of good and evil behavior, and do not rationalize our commitment to science as an intrinsic good, then the biocultural view must be remade so that the story can have a proper moral.

Until we are able to live with the consequences of the emergent anthropological notion that our political and moral preferences are no more (and no less) than preferences, the taming of evolution will continue.

Notes

Introduction: The Darwinian Revolution?

1. That this is still the state of the art was demonstrated in a recent symposium, "Teaching Bio-cultural Anthropology," at the 1980 meetings of the American Anthropological Association. All of the participants had tried to teach genuinely integrated biocultural anthropology courses and had found the existing anthropological literature woefully inadequate to the task.
2. This whig view of the history of science has been discussed as it relates to anthropology by George Stocking, Jr., in his *Race, Culture, and Evolution* (1968).
3. Some of the major works in this school are Barnes 1977; Barnes and Shapin, eds., 1979; and Wallis, ed., 1979.
4. See, for example, Young 1971, 1973.
5. Superb examples of this tradition are to be found in various general essays, such as Geertz 1973a and 1973b.
6. There are many examples of this approach. One typical study is Murdock 1965.
7. Mayr has discussed this point at length (1982).

PART I   MAJOR WESTERN VIEWS OF NATURE

1. Humoral/Environmental Theories and the Chain of Being

1. Relevant and extremely helpful exceptions to this general indictment are Gil 1969, Glacken 1967, Laín Entralgo 1970, Lovejoy [1936] 1976, Mayr 1982, Onians [1951] 1973, and Wooster [1977] 1979.
2. For a masterful discussion of these issues, see Laín Entralgo 1961. A useful collection on this subject is Caplan, Engelhardt, and McCartney, eds., 1981.

2. Evolving Natural Categories: Darwin's Unique Legacy

1. Darwin's books, letters, and notebooks are the place to begin; the secondary scholarship should be read against a thorough knowledge of Darwin's own writings. Among reviews and interpretations of Darwin, I found Eiseley 1958, Ghiselin 1969, Hull, comp. 1973, Hyman 1962, Irvine 1956, Manier 1978 particularly interesting. Gruber's publication of and commentary on Darwin's notebooks (1974) make an invaluable contribution to our understanding of the development of Darwin's thought.

   Allen 1975 provides a brief summary of the post-Darwinian modifications and developments of the theory of evolution. Hull 1974 and Beckner 1968 provide useful analytical insights into the structure and formal requirements of evolutionary theory. R. Smith 1972 gives a brief and extremely clear portrait of Wallace's views.

   On contemporary issues of major importance in the development of evolutionary theory there is an immense amount of commentary. Here I list only those I found directly useful in preparation of this work. On the phenotype/genotype distinction and associated analytical problems, Lewontin 1974 and [1974] 1976 are superb.

   Regarding the problems of units and species in evolution, in addition to Beckner 1968, Simpson [1949] 1965 and Mayr 1963 are very helpful. On the questions of the constancy of rates of selection and evolution, Williams 1966 and Gould 1977 are thought-provoking.

   The critique of theories of optimization in evolution is carried forward convincingly in Lewontin 1978 and Gould and Lewontin 1979.

   Nearly all of the works mentioned so far deal in one way or another with the problem of teleology and advance in evolution. Specific discussions of these problems are found in Mayr 1976 and Slobodkin 1977.
2. This is hardly a novel assertion about evolutionism. Yet as Ernest Mayr has recently pointed out (1982), the full implications of this reorientation have not been sufficiently appreciated. Mayr's contrast between "essentialist" and "population" thinking captures the fundamental distinction between evolutionary and pre- or nonevolutionary views. Mayr also recognizes the persistence of essentialist thinking as a dilemma in the modern biological sciences.

   Though Mayr has succeeded in pointing out that the very core of evolutionism is a revised attitude toward classification, he remains puzzled by the persistence of essentialist views in modern biology. This puzzlement leads him to explore religious and other such factors that may impinge on biological thinking and thus account for the problem. This analytical approach to the problem is not adequate; it leaves key cultural issues untouched.

3. Douglas 1966, 1973; Leach 1958, 1964, 1976; Lévi-Strauss 1962 and Schneider 1968 are representative of a much larger literature.

4. While the anthropological literature on classification is enormous, the specific relationship between classification and morality has not received sufficient explicit attention.

Part II   Simple Continuities

3. Humoral Politics: Races, Constitutional Types, and Ethnic and National Character

1. See Bateson 1979; Birdwhistell 1972; Blacking, ed., 1977.

2. See Haley 1978 and Sontag 1977–78, among many other examples.

3. See Gibson, ed., 1971 for a superb collection of texts on this subject.

Part III   Complex Continuities

1. A major exception of this stricture is Stephen Jay Gould's superb contextualization of Cyril Burtt (1981).

2. Both examples are drawn from Spanish sources. They are based on my own research and reflect my interest in the historical anthropology of Spain. Though it might seem that the choice of Spanish examples limits the generalizability of the results of my analysis, I do not think it does. At this level of generality, the Spanish case is indistinguishable from other European examples. The system of nobility in Spain was typical of Europe during the period, and the views of Spanish Enlightenment thinkers are virtually indistinguishable from those of their European contemporaries. All historical cases are unique, but nothing in the materials used in Chapters 4 and 5 could not be said on the basis of materials from other European countries.

4. Purity of Blood and Social Hierarchy

A version of this chapter was presented to the combined colloquium of the West European Studies Program and the Department of Anthropology at Indiana University. My thanks to Jerome Mintz for the invitation. It was subsequently published in Spanish in an indescribably mutilated form (Greenwood 1978). I thank Steven Kaplan, Bernd Lambert, Edmund Leach, and Dennis MacGilvary for their commentaries and suggestions for revision. James

Boon provided useful bibliographic suggestions. My initiation into the handling of these documents was provided by Julio Caro Baroja.

1. A useful discussion of these ideas is found in Onians [1951] 1973. Through analysis and comparison of multiple texts, Onians develops a portrait of Greek and Roman views on these subjects.
2. See, for example, Sabuco 1587.
3. I have written about Basque ideas on collective nobility at length elsewhere, specifically in relation to Basque ethnogenesis (Greenwood 1976a, 1977). More of the historical details can be found there. The specifics of this discussion will be limited to Guipúzcoa in order to make use of a single set of legal codes. Similar arguments can be made for Vizcaya and for parts of Alava and Navarra.
4. All translations are mine.

5. An Enlightenment Humoralist: Don Diego de Torres Villarroel

1. This point was made to me by Pilar Fernández-Cañadas de Greenwood and has been elaborated in her forthcoming paper "Los médicos del 'Canto de Calíope.'"
2. The works that have remained regularly in print are Torres Villarroel [1966] 1976 and 1972. Two others (1977 and 1979) have been republished for the first time since the 1794–99 edition of the complete works.
3. The criteria used by literary critics and historians of literature for evaluating texts differ from those of anthropologists and social historians. A text of little literary merit may be of immense anthropological value and a great work of literature can often be of little use to the social historian.
4. Partly because of the heterogeneity of subjects and literary forms and partly because of his heavyhanded satirical style, Torres has received relatively little critical attention. The most rewarding synthetic analysis of Torres' works are Granjel 1968 and Pérez' excellent introduction to *Los desahuciados del mundo y de la gloria* (1979). García Boiza's biography (1949) lacks critical depth. Mercadier's introduction to *Vida* (1972) and Sebold's introduction to *Visiones y visitas . . .* ([1966] 1976) both focus strongly on literary issues, in particular the tension in Torres between contradictory points of view. Granjel 1968 (limited to Torres' medical works) provides a good point of departure for the study of Torres.
5. All translations are mine.
6. This point is made by Valles in his introduction to *Recetarios astrológicos y alquímicos* (1977).
7. This was the only work by Torres to be sequestered by the Inquisition. Strangely, the action was taken long after the book's publication, and immediately afterward *Vida natural* was republished.

6. Human Sociobiology

The central argument and some of the materials for this chapter appeared in different form in my paper "Sociobiology: From Darwinism to Moralism," *Grinnell Magazine* 14 (1982):15–19.
1. See Barnett 1953.
2. See Greenwood and Stini 1977, chap. 20, for a critique.

Bibliography

Alexander, Richard
  1979   *Darwinism and Human Affairs*. Seattle: University of Washington Press.
Alfonso X (El Sabio)
  [c. 1265] 1848   *El código de las siete partidas*. In *Los códigos españoles*. Madrid: Rivadeneyra.
Ann Arbor Science for the People Editorial Collective
  1977   *Biology as a Social Weapon*. Minneapolis: Burgess.
Ardrey, Robert
  1966   *The Territorial Imperative*. New York: Atheneum.
Barash, David
  1979   *The Whisperings Within*. Harmondsworth: Penguin.
Barnes, Barry
  1977   *Interests and the Growth of Knowledge*. London: Routledge & Kegan Paul.
Barnes, Barry, and Steven Shapin
  1979   "Darwin and Social Darwinism: Purity and History." In *Natural Order*, ed. Barnes and Shapin, pp. 125–42. Beverly Hills: Sage.
Barnes, Barry, and Steven Shapin, eds.
  1979   *Natural Order*. Beverly Hills: Sage.

Barnett, H. G.
  1953   *Innovation*. New York: McGraw-Hill.
Bateson, Gregory
  1979   *Mind and Nature: A Necessary Unity*. New York: E. P. Dutton.
Beckner, Morton
  1968   *The Biological Way of Thought*. Berkeley: University of California
         Press.
Bennett, John
  1976   *The Ecological Transition*. New York: Pergamon.
Berlin, Brent, and Paul Kay
  1969   *Basic Color Terms*. Berkeley: University of California Press.
Birdwhistell, Ray
  1972   *Kinesics and Context*. New York: Ballantine.
Blacking, John, ed.
  1977   *The Anthropology of the Body*. London: Academic Press.
Bodin, Jean
  [1576] 1955   *Six Books of the Commonwealth*. Trans. M. J. Tooley. Ox-
         ford: Basil Blackwell.
  [1583] 1945   *Method for the Easy Comprehension of History*. Trans. B.
         Reynolds. New York: Columbia University Press.
Burton, Robert
  [1621] 1977   *The Anatomy of Melancholy: What It Is, with All the Kinds,
         Causes, Symptomes, Prognostickes & Several Cures of It*. New York:
         Vintage Books.
Caplan, Arthur; H. Tristam Engelhardt, Jr.; and James McCartney, eds.
  1981   *Concepts of Health and Disease*. Reading, Mass.: Addison-Wesley.
Caro Baroja, Julio
  1966   "Honour and Shame: An Historical Account of Several Conflicts."
         In *Honour and Shame: The Values of Mediterranean Society*, ed.
         J. G. Peristiany, pp. 79–137. Chicago: University of Chicago Press.
  1970   *El mito del carácter nacional: Meditaciones a contrapelo*. Madrid:
         Seminarios y Ediciones.
  1978   *Los judíos en la España moderna y contemporánea*. Madrid: Istmo.
Chase, Allan
  1975–76   *The Legacy of Malthus*. New York: Knopf.
Darwin, Charles
  [1859] 1958   *The Origin of Species*. New York: New American Library.
         Reprint of 6th ed.
  [1871] 1974   *The Descent of Man and Selection in Relation to Sex*. Rev.
         ed. Detroit: Gale. This ed. first published 1874.

1981   *The Descent of Man and Selection in Relation to Sex.* Princeton: Princeton University Press.

Dawkins, Richard
1976   *The Selfish Gene.* New York: Oxford University Press.

Douglas, Mary
1966   *Purity and Danger.* Harmondsworth: Penguin.
1973   *Natural Symbols.* 2d ed. New York: Vintage.

Dumont, Louis
1977   *From Mandeville to Marx.* Chicago: Univesity of Chicago Press.

Durham, William
forthcoming   *Coevolution: Genes, Culture, and Human Diversity.* Stanford: Stanford University Press.

Echave, Balthasar de
[1607] 1971   *Discursos de la antigüedad de la lengua cantabra bascongada.* Facsimile ed. Bilbao: Gran Enciclopedia Vasca.

Eiseley, Loren
1958   *Darwin's Century* Garden City, N.Y.: Doubleday.

Feijóo, Padre Benito Jerónimo de
1724–39   *Theatro crítico universal ó Discursos Varios, en Todo Genero de Materias, Para Desengaño de Errores Comunes.* 8 vols. Madrid: Lorenzo Francisco Mojados.

Freeman, Derek
1982   *Margaret Mead and Samoa.* Cambridge: Harvard University Press.

Galton, Francis
[1869] 1962   *Hereditary Genius.* Cleveland: World.

García Boiza, Antonio
1949   *Don Diego de Torres Villarroel: Ensayo biográfico.* Madrid: Editora Nacional.

Geertz, Clifford
1973a   "The Growth of Culture and the Evolution of Mind." In Geertz, *The Interpretation of Cultures*, pp. 55–83. New York: Basic Books. First published 1962.
1973b   "The Impact of the Concept of Culture on the Concept of Man." In Geertz, *The Interpretation of Cultures*, pp. 33–54. New York: Basic Books. First published 1966.

Ghiselin, Michael
1969   *The Triumph of the Darwinian Method.* Berkeley: University of California Press.

Gibson, Charles, ed.
1971   *The Black Legend.* New York: Knopf.

Gil, Luis
　1969　*Therapeia: La medicina popular en el mundo clásico.* Madrid: Guadarrama.

Glacken, Clarence
　1967　*Traces on the Rhodian Shore.* Berkeley: University of California Press.

Gould, Stephen Jay
　1977　*Ever since Darwin.* New York: W. W. Norton.
　1981　*The Mismeasure of Man.* New York: W. W. Norton.

Gould, Stephen Jay, and Niles Eldredge
　1977　"Punctuated Equilibria: The Tempo and Mode of Evolution Reconsidered." *Paleobiology* 3:115–51.

Gould, Stephen Jay, and Richard Lewontin
　1979　"The Spandrels of San Marco and the Panglossian Paradigm: A Critique of the Adaptationist Programme." *Proceedings of the Royal Society of London* B205:581–98.

Granjel, Luis S.
　1968　"La medicina y los médicos en las obras de Torres Villarroel." In Granjel, *Humanismo y medicina,* pp. 247–313. Estudios de Historia de la Medicina Española, vol. 2. Salamanca: Universidad de Salamanca.

Greenwood, Davydd
　1976a　"Ethnic Regionalisms in the Spanish Basque Country." *Iberian Studies* 5:49–52.
　1976b　*Unrewarding Wealth: Commercialization and the Collapse of Agriculture in a Spanish Basque Town.* Cambridge: Cambridge University Press.
　1977　"Continuity in Change: Spanish Basque Ethnic Identity as an Historical Process." In *Ethnic Conflict in the Western World;* ed. Milton Esman, pp. 81–102. Ithaca: Cornell University Press.
　1978　"Pureza de sangre y nobleza en el País Vasco y Castilla: Conceptos naturalistas, variedades del orden social, y autonomía de la naturaleza." *Ethnica* 14:163–82.
　1982　"Sociobiology: From Darwinism to Moralism." *Grinnell Magazine* 14:15–19.

Greenwood, Davydd, and William Stini
　1977　*Nature, Culture, and Human History: A Biocultural Introduction to Anthropology.* New York: Harper & Row.

Greenwood, Pilar F.-C.
　forthcoming　"Los médicos del 'Canto de Calíope.'" *Quaderni Ibero-Americani.*

Gruber, Howard, ed.
1974 *Darwin on Man*. London: Wildwood House.

Haley, Bruce
1978 *The Healthy Body and Victorian Culture*. Cambridge: Harvard University Press.

Hamilton, W. D.
1964 "The Genetical Evolution of Social Behavior, I and II." *Journal of Theoretical Biology* 7:1–52.
1970 "Selfish and Spiteful Behaviour in an Evolutionary Model." *Nature* 228 (5277):1218–20.
1971a "Geometry for the Selfish Herd." *Journal of Theoretical Biology* 31:295–311.
1971b "Selection of Selfish and Altruistic Behavior in Some Extreme Models." In *Man and Beast*, ed. J. Eisenberg and W. Dillon, pp. 57–91. Washington, D.C.: Smithsonian Institution Press.

Harris, Marvin
1964 *The Nature of Cultural Things*. New York: Random House.
1968 *The Rise of Anthropological Theory*. New York: Crowell.
1971 *Culture, Man, and Nature*. New York: Crowell.
[1974] 1975 *Cows, Pigs, Wars, and Witches*. New York: Vintage.
1977 *Cannibals and Kings: The Origins of Cultures*. New York: Random House.
1979 *Cultural Materialism*. New York: Random House.
1981 *America Now: The Anthropology of a Changing Culture*. New York: Simon & Schuster.

Herbert, Sandra, ed.
1980 *The Red Notebook of Charles Darwin*. Ithaca: Cornell University Press.

Hippocrates
1886 "Airs, Waters, and Places." In *The Genuine Works of Hippocrates*, vol. 1, ed. F. Adams. New York: William Wood.

Hull, David
1974 *The Philosophy of Biology Science*. Englewood Cliffs, N.J.: Prentice-Hall.

Hull, David, comp.
1973 *Darwin and His Critics*. Cambridge: Harvard University Press.

Hyman, Stanley
1962 *The Tangled Bank*. New York: Atheneum.

Irvine, William
1956 *Apes, Angels, and Victorians*. London: Weidenfeld & Nicholson.

Isasti, Lope de
[1625, 1850] 1972   *Compendio historial de la M.N. y M.L. Provincia de Guipúzcoa.* Fascimile ed. of 1850 ed. Bilbao: Gran Enciclopedia Vasca.

Jacob, François
1982   *The Possible and the Actual.* New York: Pantheon.

Kretschmer, Ernst
1925   *Physique and Character.* Trans. W. J. H. Sprott. 2d rev. ed. New York: Harcourt, Brace. First English ed. published 1921.

Laín Entralgo, Pedro
1961   *Enfermedad y pecado.* Barcelona: Toray.
1970   *La medicina hipocrática.* Madrid: Revista de Occidente.

Larramendi, Manuel de, S.J.
[1754] 1969   *Corografía o descripción general de la Muy Noble y Muy Leal Provincia de Guipúzcoa.* San Sebastián: Sociedad Guipuzcoana.

la Souchère, Eléna de
1964   *An Explanation of Spain.* Trans. Eleanor Ross Levieux. New York: Random House.

Leach, E. R.
1958   "Magical Hair." *Journal of the Royal Anthropological Institute* 88:147–64.
1964   "Anthropological Aspects of Language: Animal Categories and Verbal Abuse." In *New Directions in the Study of Language,* ed. E. H. Lenneberg. Cambridge: M.I.T. Press.
1976   *Culture and Communication.* Cambridge: Cambridge University Press.

Lee, Richard, and Irven De Vore, eds.
1968   *Man, the Hunter.* Chicago: Aldine.

Lévi-Strauss, Claude
1962   *Totemism.* Trans. R. Needham. Boston: Beacon Press.
1963   *Structural Anthropology.* New York: Basic Books.

Lewontin, Richard
1974   *The Genetic Basis of Evolutionary Change.* New York: Columbia University Press.
[1974] 1976   "The Analysis of Variance and the Analysis of Causes." In *The IQ Controversy,* ed. N. Block and G. Dworkin, pp. 179–92. New York: Pantheon.
1978   "Adaptation." *Scientific American* 239:156–69.

Lorenz, Konrad
   [1963] 1971   *On Aggression.* Trans. Marjorie Kerr Wilson. New York:
      Bantam Books.

Lovejoy, Arthur
   [1936] 1976   *The Great Chain of Being.* Cambridge: Harvard University
      Press.

Lumsden, Charles, and E. O. Wilson
   1981   *Genes, Mind, and Culture.* Cambridge: Harvard University Press.

Manier, Edward
   1978   *The Young Darwin and His Cultural Circle.* Dordrecht: Reidel.

Martínez, Martín
   1748   *Medicina scéptica y cirugía moderna* . . . 2 vols. 3d ed. Madrid:
      Imprenta Real.

Mayr, Ernst
   1963   *Animal Species and Evolution.* Cambridge: Harvard University
      Press.
   1976   *Evolution and the Diversity of Life.* Cambridge: Belknap Press of
      Harvard University Press.
   1982   *The Growth of Biological Thought.* Cambridge: Harvard Univer-
      sity Press.

Mead, Margaret
   1928   *Coming of Age in Samoa.* New York: William Morrow.

Midgley, Mary
   1978   *Beast and Man.* Ithaca: Cornell University Press.

Moreno de Vargas, Bernabé
   [1636] 1795   *Discursos de la nobleza de España.* Madrid: Antonio Espi-
      nosa.

Morris, Desmond
   1969   *The Human Zoo.* New York: Delta.

Murdock, George Peter
   1965   "The Common Denominator of Cultures." In Murdock, *Culture
      and Society,* pp. 88–110. Pittsburgh: University of Pittsburgh Press.

Netting, Robert McC.
   1971   *The Ecological Approach to Cultural Study.* Reading, Mass.: Ad-
      dison-Wesley.

*Nueva recopilación de los fueros, privilegios, buenos usos y costumbres, leyes
y órdenes de la M.N. y M.L. Provincia de Guipúzcoa*
   [1696] 1918   San Sebastián: Imprenta de la Provincia.

Onians, Richard
  [1951] 1973 *The Origins of European Thought about the Body, the Mind, the Soul, and Fate.* New York: Arno Press.

Otazu y Llana, Alfonso
  1973 *El "igualitarismo" vasco: Mito y realidad.* Bilbao: Txertoa.

Pritchett, V. S.
  [1954] 1965 *The Spanish Temper.* New York: Harper & Row.

Reynolds, Vernon
  1980 *The Biology of Human Action.* 2d ed. Oxford: W. H. Freeman.

Sabuco, Miguel
  1587 *Nueva Filosofía de la naturaleza del hombre, no conocida ni alcanzada de los grandes Filósofos antiguos, la cual mejora la vida y salud humana.* Madrid. (Published under the name of his daughter Oliva Sabuco de Nantes.)

Sahlins, Marshall
  1976 *The Use and Abuse of Biology.* Ann Arbor: University of Michigan Press.
  1978 "Culture as Protein and Profit." *New York Review of Books,* November 23, pp. 45–53.

Schneider, David
  1968 *American Kinship: A Cultural Account.* Englewood Cliffs, N.J.: Prentice-Hall.

Sheldon, W. H. (with S. S. Stevens and W. B. Tucker)
  1940 *The Varieties of Human Physique: An Introduction to Constitutional Psychology.* New York: Harper.

Simpson, G. G.
  [1949] 1965 *The Meaning of Evolution.* New Haven: Yale University Press.

Slobodkin, Lawrence
  1977 "Evolution Is No Help." *World Archaelogy* 8:332–43.

Smith, Eric
  1980 "Evolutionary Ecology and the Analysis of Human Foraging Behavior: An Inuit Example from the East Coast of Hudson Bay." Ph.D. dissertation, Cornell University.

Smith, Roger
  1972 "Alfred Russel Wallace: Philosophy of Nature and Man." *British Journal for the History of Science* 6:177–99.

Sociobiology Study Group
  1977 "Sociobiology: A New Biological Determinism." In Ann Arbor

Science for the People Editorial Collective, *Biology as a Social Weapon*, pp. 133–49. Minneapolis: Burgess.

Sontag, Susan
1977–78   *Illness as Metaphor*. New York: Farrar, Straus & Giroux.

Spuhler, James, ed.
1959   *The Evolution of Man's Capacity for Culture*. Detroit: Wayne State University Press.

Stocking, George, Jr.
1968   *Race, Culture, and Evolution*. New York: Free Press.

Torres Villarroel, Diego de
1794–99   *Obras completas*. 15 vols. Madrid: Viuda de Ibarra.
1972   *Vida: Ascendencia, nacimiento, crianza y aventuras*. Ed. Guy Mercadier. Madrid: Clásicos Castalia. First published sequentially in 1743, 1750, 1752, and 1758.
[1966] 1976   *Visiones y visitas de Torres con Don Francisco de Quevedo por la Corte*. Ed. Russell Sebold. Madrid: Clásicos Castellanos 61. First published 1727–28.
1977   *Recitarios astrólogico y alquímico*. Ed. José Manuel Valles. Madrid: Editora Nacional. First published 1726.
1979   *Los desahuciados del mundo y de la gloria*. Ed. Manual María Pérez. Madrid: Editora Nacional. First published 1737.

Wallis, Roy, ed.
1979   *On the Margin of Science*. Monograph no. 27. Keele: Sociological Review.

Williams, G. C.
1966   *Adaptation and Group Selection*. Princeton: Princeton University Press.

Wilson, E. O.
1975   *Sociobiology: The New Synthesis*. Cambridge: Belknap Press of Harvard University Press.
1978   *On Human Nature*. Cambridge: Harvard University Press.

Winterhalder, Bruce
1977   "Foraging Strategy Alternatives of the Boreal Forest Cree." Ph.D. dissertation, Cornell University.

Wooster, Donald
[1977] 1979   *Nature's Economy*. Garden City, N.Y.: Anchor Books.

Wynne-Edwards, V. C.
1962   *Animal Dispersion in Relation to Social Behaviour*. Edinburgh: Oliver & Boyd.

Young, R. M.
  1971   "Evolutionary Biology and Ideology: Then and Now." *Science Studies* 1:177–206.
  1973   "The Historiographic and Ideological Contexts of the Nineteenth-Century Debate on Man's Place in Nature." In *Changing Perspectives in the History of Science*, ed. M. Teich and R. M. Young, pp. 344–438. London: Heinemann.

Zaldivia, Bachiller Juan Martínez de
  [1517] 1944   *Suma de las cosas cantbricas y guipuzcoanas.* San Sebastián: Excma. Diputación de Guipúzcoa.

Index

Library of Congress Cataloging in Publication Data

GREENWOOD, DAVYDD J.
    The taming of evolution.

    Bibliography: p.
    Includes index.
    1. Human evolution.   2. Nature and nurture.
3. Sociobiology.   4. Physical anthropology—
Philosophy.   I. Title.
GN281.G73   1984   573   84-45147
ISBN 0-8014-1743-0   (alk. paper)